ENERGY POLICIES, POLITICS AND PRICES

GEOTHERMAL EXPLORATION AND DEVELOPMENT POLICY

INTERNATIONAL LESSONS

ENERGY POLICIES, POLITICS AND PRICES

Additional books in this series can be found on Nova's website under the Series tab.

Additional e-books in this series can be found on Nova's website under the e-book tab.

ENERGY POLICIES, POLITICS AND PRICES

GEOTHERMAL EXPLORATION AND DEVELOPMENT POLICY

INTERNATIONAL LESSONS

LEON EDWARDS
EDITOR

nova publishers
New York

For permission to use material from this book please contact us:
Telephone 631-231-7269; Fax 631-231-8175
Web Site: http://www.novapublishers.com

Library of Congress Cataloging-in-Publication Data

ISBN: 978-1-63321-824-6

Published by Nova Science Publishers, Inc. † New York

CONTENTS

PREFACE

Chapter 1 - This report focuses on five of the policy types that are most relevant to the U.S. market and political context for the exploration and confirmation of conventional hydrothermal (geothermal) resources in the United States: (1) drilling failure insurance, (2) loan guarantees, (3) subsidized loans, (4) capital subsidies, and (5) government-led exploration. It describes each policy type and its application in other countries and regions. It offers policymakers a guide for drafting future geothermal support mechanisms for the exploration-drilling phase of geothermal development.

Chapter 2 - Developing power plants is a risky endeavor, whether conventional or renewable generation. Feed-in tariff (FIT) policies can be designed to address some of these risks, and their design can be tailored to geothermal electric plant development. Geothermal projects face risks similar to other generation project development, including finding buyers for power, ensuring adequate transmission capacity, competing to supply electricity and/or renewable energy certificates (RECs), securing reliable revenue streams, navigating the legal issues related to project development, and reacting to changes in existing regulations or incentives. Although FITs have not been created specifically for geothermal in the United States to date, a variety of FIT design options could reduce geothermal power plant development risks and are explored. This analysis focuses on the design of FIT incentive policies for geothermal electric projects and how FITs can be used to reduce risks (excluding drilling unproductive exploratory wells).

In: Geothermal Exploration …
Editor: Leon Edwards

ISBN: 978-1-63321-824-6

Chapter 1

Geothermal Exploration Policy Mechanisms: Lessons for the United States from International Applications*

Bethany Speer, Ryan Economy, Travis Lowder, Paul Schwabe and Scott Regenthal

List of Acronyms

ADEME	Environment and Energy Management Agency (France)
AIST	Agency of Industrial Science and Technology (Japan)
DOE	U.S. Department of Energy
FIPP	Financial Institution Partnership Program
GEA	Geothermal Energy Association
GFF	Geothermal Fund Facility
GW	gigawatts (electric)
ITC	investment tax credit
KfW	Kreditanstalt für Wiederaufbau Bankengruppe of Germany
MOL	Hungarian Oil and Gas Company
MW	megawatts (electric)

* This is an edited, reformatted and augmented version of technical report, NREL/TP-6A20-61477, issued by the National Renewable Energy Laboratory, May 2014.

NEDO	New Energy Development Organization (Japan)
PTC	production tax credit
USGS	United States Geological Survey

EXECUTIVE SUMMARY

According to a U.S. Geological Survey (USGS) estimate, the United States has nearly 40 gigawatts (GW) of power generation potential from identified and unidentified conventional geothermal resources (Williams et al. 2008a).[1] To realize these resources, geothermal project developers must overcome several obstacles that are unique among the renewable energy technologies. One significant barrier in geothermal project development is the high investment risk during the resource exploration phase, which can make financing a geothermal project difficult as compared to other renewable energy sources, including wind and solar (Salmon et al. 2011).

Many federal and state policies provide incentives to renewable energy sources, including geothermal; however, these policies rarely differentiate between the technologies, with the common exception of the level of remuneration provided (e.g., the amount of a rebate or tax credit). These incentives, therefore, may not adequately address the more nuanced support required to advance geothermal technologies (Doris et al. 2009). For example, at present many geothermal policies support the operational phase of the project, but much of the risk is in the development phase. If policy-makers wish to incentivize development of geothermal power capacity, policies may need to address exploration risks specifically, thereby improving developers' access to financing through this vital stage in the development cycle.

A number of governments (both in the United States and abroad) and international development organizations have supported exploration and confirmation of conventional hydrothermal geothermal resources with a variety of public policies. In order to define the scope of this analysis and provide conceptual clarity, this report focuses on the five general policy types listed below, as well as hybrid combinations that have been applied to geothermal and are understood to have the potential to support geothermal power in the United States, specifically (see Table ES-1):

1. Loan guarantees
2. Drilling failure insurance
3. Lending support mechanisms

4. Grants
5. Government-led exploration.

The analysis describes each policy type applicable to the exploration drilling phase and presents examples of their use in a variety of countries and regions. It also assesses each policy's potential applicability to the U.S. geothermal market. The report is intended to offer a review of select past and current policy applications to policymakers, providing a basis for evaluating options for future geothermal exploration policies.

Table ES-1. Policy Descriptions, Examples, Summary of Characteristics, and Potential Applicability to the United States

Policy Description	Examples	Characteristics[a]	Potential Applicability to the U.S. Market
Loan Guarantee: As a third party to the transaction, a government agency or other public entity provides a guarantee of debt repayment to a lender in the event of borrower default. A fee is often required of the developer.	United States; Germany	Loan guarantees can provide high leverage of private investment in case of low payouts but have thus far demonstrated limited results in the U.S. geothermal market.	The Section 1705 Loan Program within the U.S. Department of Energy (DOE) provided \$545.5 million in guarantees to five geothermal projects, although similar programs for renewable energy projects may be unlikely due to the political environment following bankruptcies of a few loan guarantee recipients; however, these guarantees were provided to manufacturers and not generators. Designing loan guarantees to cover loans provided specifically for the exploration phase could spur more debt financing at a phase primarily funded with more expensive equity.
Drilling Failure Insurance: The developer pays a premium to access a government-provided partial-cost reimbursement. The reimbursement is paid in the event	France; Multilateral banks and development agencies	Drilling failure insurance can provide high leverage of private investment when the number of claims is limited. Premiums may be cost-prohibitive for developers, and there is the potential for large public funding expenditures depending	Policymakers could offer drilling failure insurance as an alternative to loan guarantees or lending support mechanisms to reduce the cost of financing. An insurance program would require a significant up-front allocation of funds to capitalize the insurance pool.

Table ES-1. (Continued)

Policy Description	Examples	Characteristics[a]	Potential Applicability to the U.S. Market
that the exploration phase does not result in a predetermined level of success for a given project.		on the total value of the claims paid.	
Lending Support Mechanisms: A government entity reduces the cost of the loan by buying down the interest rate, providing for a longer loan term, paying a portion of the interest, or offering some other loan support.	Germany	In instances when the terms required by lenders may be prohibitive for the developer, government interest rate support could make loan terms more amenable (p. 37). Government-supported loans (e.g., direct loans or public-private partnerships) could successfully leverage private investment at subsequent stages of development but may crowd out private investment in the exploration phase.	An interest rate subsidy provided by the government to a private lender could be less costly to the government than outright direct loans, but such a program would still require significant public investment. Public investment would also be unrecoverable unless developers are required to pay back the interest subsidy once the resource is proven.
Grants and Cooperative Agreements (Grants): A government entity or other institution offers cost-sharing schemes or other forms of direct payments intended to reduce the investment cost during the exploration phase.	United States; France; Multilateral banks and development agencies; Australia; Iceland; Japan	Grants and cooperative agreements can be effective for individual project development, but depending on the amount of support provided, they may result in limited leverage of private investment during the exploration phase, leading to significant public funding commitments.	Grants and cooperative agreements successfully spurred initial growth of geothermal projects in the United States but may be limited in their ability to significantly impact overall market development today due to the low leverage of private investment and higher funding requirements.
Government-Led Exploration: A government entity undertakes exploration activities directly or contracts private firms to do so on their behalf. Proven resources are	Iceland; Japan; New Zealand; Indonesia	Shown to be effective for project development in markets where little to no exploration activity is occurring. However, government-led exploration provides lowest leverage of private investment and	Despite historical federal support of geothermal exploration, this policy may not be well suited to the United States at present. This is because it is not market oriented and because of the perception that the geothermal market is more mature than it was when

Policy Description	Examples	Characteristics[a]	Potential Applicability to the U.S. Market
developed by government-owned enterprises or auctioned to private firms.		highest level of public investment.	first spurred in part by the government in the 1970s and 1980s.

[a] The organizing principle of this section is "leverage" or the level to which a policy can incentivize private investment with public funds The assessments of leverage provided within the characteristics column are general comparisons across the five policy types. Actual leverage will depend on the specifics of a policy's design.

The key takeaways from this analysis include:

- **If policymakers want to encourage more geothermal exploration, they could consider policies that initiate lending to early stage projects.** Nearly all U.S. projects are funded by equity during the exploration phase. Encouraging more lending during the exploration phase would provide another form of capital in the form of debt and could be done via loan guarantees, government direct loans, or a combination of both. The advantage of accessing greater proportions of debt is that this could help lower a project's weighted cost of capital due to the typically lower cost of borrowing funds as compared to gaining equity investors. Encouraging lenders to participate in the exploration phase or providing loans directly via a "Green Bank" or a program similar to the Loans for Geothermal Reservoir Confirmation program could help reduce the cost of capital and increase the volume of funding available at the exploration phase.
- **There are policy options that help encourage additional equity investment during the exploration phase.** For example, re-authorizing a cost-share grant program, such as those utilized in the United States in the 1970s and 1980s, could help developers limit the risk to investors during the exploration phase.
- **Geothermal-specific policy innovations can account for unique project development characteristics.** One possibility could be tax incentives for exploration-phase drilling. Exploration-drilling tax incentives—either in the form of credits or deductions—could decrease geothermal development companies' tax expenses during the exploration phase, offsetting some of the financial risk incurred during the exploration phase. Other policy options include incentives to encourage the assessment or utilization of geothermal fluids co-

produced from oil and gas exploration, which could help expand the resource base.

- **Some geothermal incentive policies do not fit neatly into clearly defined categories.** Several countries and multilateral agencies have implemented or are developing support mechanisms that overlap the five conceptual policies summarized above. Exploration policies can be creatively designed and can be a hybrid of the five general policies types described in this report. Examples of this include the loan/cost-share drilling program in Japan, the Risk of Non-Discovery of Deep Geothermal Energy loan program in Germany, and the Geothermal Fund Facility in Indonesia.

1. Introduction

The U.S. geothermal market is the largest in the world, with an installed capacity of 3,386 megawatts (MW) as of February 2013 (GEA 2013). However, the pace of new utility-scale geothermal electric projects coming online in the United States has been slow over the last two decades. Even considering the most recent period of robust government investment for renewables following passage of the American Recovery and Reinvestment Act of 2009 (Recovery Act), the U.S. market has only grown about 14% in the decade preceding 2012 (EERE 2012).

Despite providing reliable baseload power and being a commercially viable technology, conventional hydrothermal generation projects face several key challenges to development:

- A general lack of investors familiar with the technology and appropriate deal structures (Salmon et al. 2011)
- Long lead times between project conception and completion (typically four to five years)
- Remote locations of project sites, potentially requiring expensive build-outs of transmission capacity that add to project timelines and costs (Hurlbut 2012)
- Competition for capital and equipment with other extraction industries, including the coal and oil and gas industries (Salmon et al. 2011)
- High risks and subsequent costs of capital during exploration drilling.

In this report, we focus on the last barrier because the high risks and cost of capital associated with the exploration phases is a key challenge to geothermal plant development (Salmon et al. 2011). Failure to confirm the geothermal resource during the exploration phase results in a forfeiture of invested capital. This risk might drive investors away from geothermal financing, especially those lenders who have risk appetites that are typically lower than that of equity investors.[2] Further, those investors are likely to require higher yields to compensate for the risk as compared to other stages of geothermal plant development.

The primary difference between developing a geothermal project compared to other renewable energy technologies lies in the challenges of identifying and confirming the resource (Salmon et al. 2011). With geothermal projects, the resources must be identified and confirmed on a site- specific basis whereas with other technologies, such as wind and solar, the initial identification can be done on a regional basis (although further assessment of local resources may be necessary, especially with utility-scale projects). However, current renewable energy policy in the United States does not address geothermal exploration risks specifically (i.e., the production tax credit). In addition, current renewable energy policies provide operational incentives, which do not directly support these pre-operational and yet vital—and costly—phases of geothermal power plant development.

This analysis focuses on policy options that could potentially fill this gap by providing geothermal developers with pre-operational incentives to target the initial, high-risk project development phases. It builds on the following geothermal policy analyses by referencing relevant research where pertinent, while avoiding repetition of the previously conducted work: Speer (2012), Rickerson et al. (2012), Salmon et al. (2011), and Doris et al. (2009).

1.1. Background

The risk profile for geothermal projects is unlike other renewable energy technologies, such as wind and solar, that can quantify their resources through comparatively inexpensive site assessments and accessible weather data. In contrast, geothermal projects can only verify the existence and quality of the resource through exploration drilling, which is expensive and may not result in a feasible project or one that produces as much power as originally estimated. Although the International Finance Corporation estimates that 59% of wells drilled during the exploration phase have been successful globally, the rate in the United States for well-researched and vetted sites may be in the range 35%–50% when economic and other factors are taken into consideration (IFC 2013; Speer 2012).[3]

It is also possible to leverage oil and gas drilling results to help identify geothermal resources. Some oil and gas projects may also be viable for geothermal co-production; however, co- production projects are currently in the demonstration phase and may have more limited applications due to their remote locations and smaller capacity as compared to utility-scale hydrothermal plants. At the time of writing, ElectraTherm had commissioned the only co- production geothermal power plant in the United States (GEA 2013).

There is no singular definition of the phases for developing geothermal projects. For the purposes of this report, we chose to use the Geothermal Energy Association's (GEA) description of the phases, which are outlined in Figure 1 (GEA 2010a). This report focuses on the first and second phases, which are collectively referred to herein as "exploration."

The outcome of Phases I and II determines the future of the project. The full range of activities required to confirm a potential geothermal resource, including drilling a single full-sized discovery well to determine the presence and quality of the geothermal resource, can cost $6 million to $10 million (Project Finance 2010).

The probability that the well will be dry or partially dry (i.e., the temperature and flow-rate are too low to meet generation targets) is considerable (Salmon et al. 2011). If the resource cannot be proven or it is otherwise insufficient for the project's needs, the project must be abandoned and investor capital is lost.

The challenges of the exploration phases are exemplified by the consistently larger number of projects in Phases I and II starting in 2010 compared to the number of projects in later phases of development (e.g., Phase III and beyond) (see Figure 2.) This backlog of projects could be due in part to the exploration risk and subsequent financing challenges that may be preventing the projects reaching the next stage of development (NREL analysis; GEA 2013).

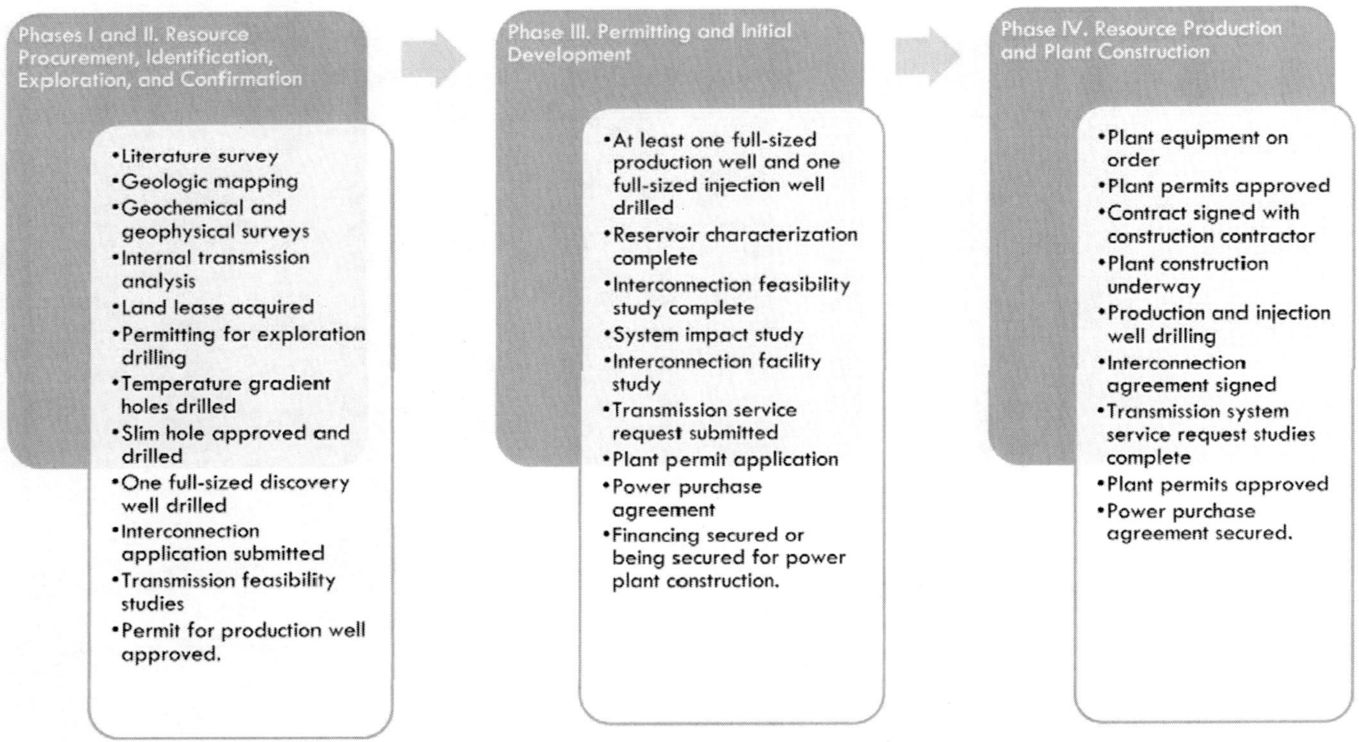

Figure 1. U.S. Geothermal Energy Association's terms and definitions for geothermal resource development (2010a).

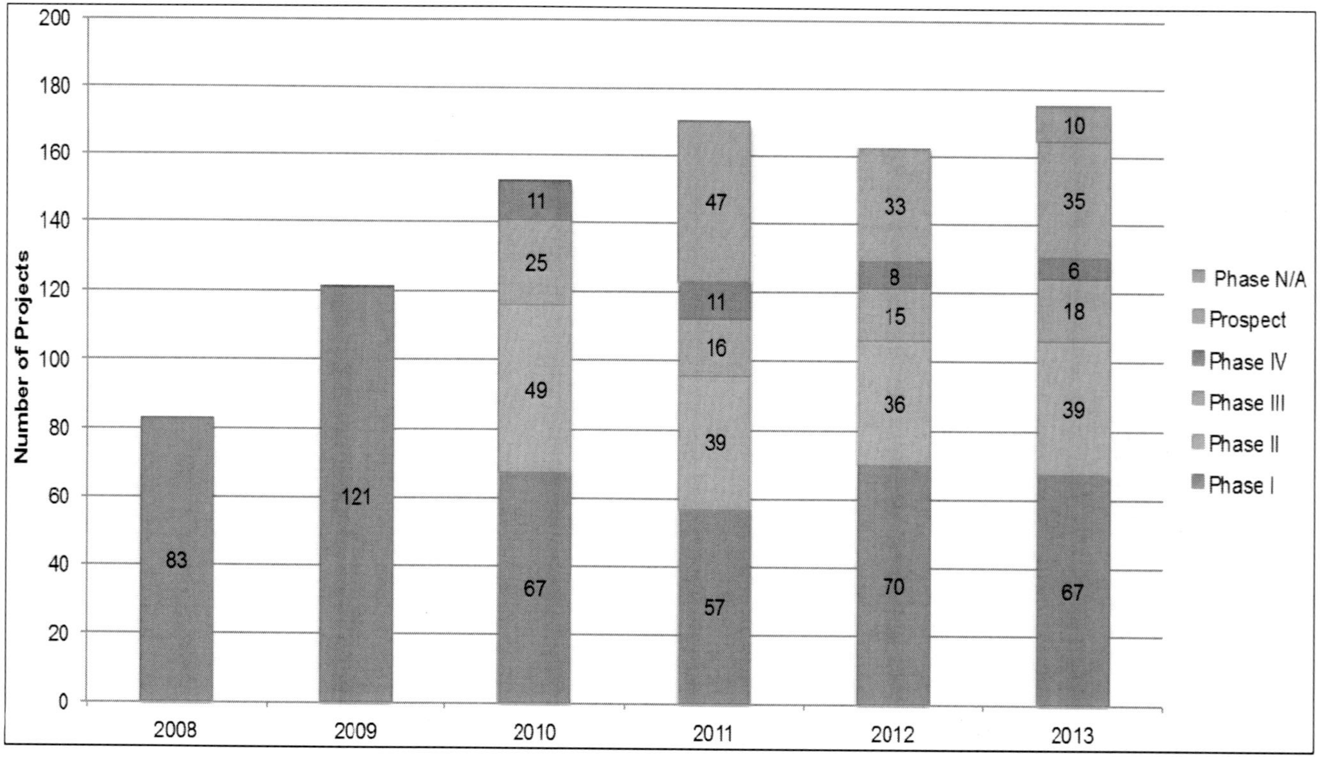

Figure 2. Total projects by year and phase (including unconfirmed projects) (GEA 2013).[4]

1.2. Purpose and Scope

This report focuses on hydrothermal technologies as they are the most commercialized in the United States; we do not focus on other innovations, such as enhanced geothermal systems (EGS) and co-production. Because of their unique risk profile, hydrothermal geothermal projects benefit from different policy structures than other renewable energy technologies to support their deployment. The purpose of this analysis is to help inform policymakers of options that could be used to support geothermal exploration based on previous policy experiences, including those outside of the United States. We have focused on the five general policy types noted below because they are believed to have the most applicability to the U.S. policy context. With the exceptions of drilling insurance and government-led exploration, all other policies can support other phases of geothermal activity (not just exploration); however, these same policies can be designed to specifically encourage geothermal exploration. Note that several countries have customized policies to suit the needs of their geothermal markets and deployment goals, and in doing so, have created hybridized structures that are described in greater detail in the country sections included later in this report.

- **Loan guarantee**—A government or other public entity provides a guarantee of debt repayment to a lender in the event of borrower default; a loan guarantee does not provide for a loan itself.
- **Drilling failure insurance**—The developer pays a premium to access a government- provided partial-cost reimbursement in the event of a drilling failure or a less-than- expected level of success.
- **Lending support mechanism**—A government entity reduces the cost of a loan in a variety of ways. The government can work with a private lender to provide a loan program and can buy down the interest rate charged. Other options include longer interest rates, interest-free periods, and payment-free periods, all of which would likely apply only under a government-provided loan program.
- **Grants and Cooperative Agreements**—These include cost-sharing schemes or other forms of direct payment intended to reduce the investment cost during early-stage development.
- **Government-led exploration**—A government entity leads, partners, or contracts work to identify and prove new geothermal resources. The proven resource may be subsequently transferred (e.g., via a

tendering process) to a private sector entity for project development, operation, and/or ownership.

We deemed some countries' policies to be highly unlikely or impractical in the U.S. market and policy context, and thus omitted them from this report. For example, the Government of Kenya has compensated for a general lack of private investment in the country by operating a nationalized geothermal development company that is largely capitalized with aid funding (with additional funds deriving from national tax revenues).

This report also discusses three multilateral agency programs that include risk mitigation instruments whose design could be replicated by the United States. Although these programs were or are funded by donors and are intended to spur development in less-developed countries, they nonetheless provide an instructive example of how to address geothermal resource risk.

Furthermore, some of the countries discussed in this report have a very small amount of identified hydrothermal resources (e.g., Australia, Germany, and France) and have designed their policies to support geothermal heat projects and/or enhanced geothermal systems (a technology that is, as of this writing, not yet fully commercialized).[5] We have included these countries and their policies in the analysis because the policy structures are relevant to hydrothermal applications in the U.S. market.

1.3. Report Organization

Section 2 begins the analysis with an overview of the countries discussed in this report. Each country is assessed according to the following:

- Installed geothermal capacity
- Estimated geothermal resource potential
- Proportion of geothermal capacity to total installed capacity
- Sources of financing and utility regulatory context
- Current policy mechanisms supporting geothermal project development.

Section 3 looks further into the U.S. context with a brief history of its support of geothermal development at the federal level, including exploration policies.

Section 4 provides conceptual descriptions of each of the five general policy types. Here, we describe the abstracted form of these policies, separate from their practical applications. The organizing principle of this section is "leverage" or the level to which a policy can incentivize private investment with public funds.

Section 5 discusses the in-country applications of the five policy types. This section is organized by country instead of policy type (as is Section 4) because in practice, the divisions between these policies are not always clear or countries have employed multiple policies, making it difficult to categorize a country under a single policy.

Section 6 assesses the characteristics of the five policy types (according to megawatts deployed and private investment leveraged), bringing together insights from the case studies examined in Section 5 and analyzing each policy's applicability to the U.S. policy context.

Section 7 provides suggestions on innovative policies that could support geothermal exploration, concludes the report, and proposes areas for further research.

2. Country Context

Table 1 summarizes the political, economic, and geothermal market context for each of the countries discussed in this report, as well as the policy types that have been used within the countries.

It should be noted that five of the top ten countries in terms of installed geothermal capacity are not included below because we could either not identify any exploration policy regimes or those regimes were deemed not applicable to the United States due to the structure of the U.S. electric utility sector and the institutions that are available to provide incentives. These five countries not included, in order of installed capacity, are the Philippines, Mexico Italy, El Salvador, and Costa Rica (BNEF 2012). The Philippines is not included in the main analysis because the state-owned national oil company *developed* nearly all of its existing geothermal capacity. However, a description of the Philippines's geothermal development is included in Appendix A because the country recently privatized the national oil company and enacted a number of policies designed to support geothermal development along with other sources of renewable energy. Also included in Appendix B is a brief discussion of the Government of Kenya's unique approach to geothermal exploration led by a state-owned enterprise. We were unable to identify the policy regimes Italy, El Salvador, and Costa Rica. And as of the time of writing this report, Mexico was formulating but had not yet enacted a program to provide insurance and financing for geothermal exploration activities; for more information, see: https://www.climateinvestmentfunds.org/cif/sites/climateinvestmentfunds.org/files/Mexico%20Geothermal%20Risk%20Mitigation%20Facility%20-%20public.pdf.

3. Background on U.S. Federal Geothermal Power Policies Supporting All Phases of Development

The United States leads the world in installed geothermal capacity with 3,386 MW. To reach this point, national and state government agencies employed a number of supporting policies and financial incentives ranging from loan guarantees and tax credits to direct subsidies. During the 1970s, 1980s, and early 2000s, the federal government enacted a number of policies and incentives that helped jump start—and then advance—the geothermal industry. These included:

- The first federal loan guarantee program begun in 1974 (Nasr 1978; Doris et al 2009; Speer 2012). Lead agencies: Energy Research and Development Administration and later the Department of Energy
- The Public Utilities Regulatory Policies Act begun in 1978 (Owens 2002; Doris et al. 2009). Lead agency: Federal Energy Regulation Commission with some responsibilities delegated to states
- The investment tax credit (ITC) begun in 1978 (Doris et al. 2009; Speer 2012). Lead agency: Department of Treasury
- The Industry-Coupled Case Studies Program begun in 1978 (Moore et al. 2010; Speer 2012). Lead agency: Department of Energy
- The production tax credit (PTC) begun in 2005 (Doris et al. 2009; DSIRE 2012; Speer 2012). Lead agency: Department of Treasury.

Most of these programs supported renewable energy development more generally; however, federal programs specifically relevant to the exploration drilling (including the loan guarantee and Industry-Coupled programs briefly mentioned here) are presented in more detail in Section 5.1.

Table 1. Summary Information for Countries Examined Herein

Country and Installed Capacity Rank	Geothermal Information	Background	Exploration Support Type
United States 1st	**Installed Capacity** 3,386 MW **Hydrothermal Resource Potential** 39,090 MW[a] **Percentage of Total Power Grid Capacity** ~0.2%[b]	The United States has a patchwork of deregulated and regulated electricity markets in which power prices are determined by a host of market forces, regulations, and policies. Utilities and power generators can be government-owned or privately held, however most geothermal power generators are privately developed and owned and operated. The federal government has advanced geothermal project development by supporting exploration activities for hydrothermal resources in the 1970s and 1980s and has since provided a number of incentives and supporting policies. States and local governments have also supported geothermal projects; however, these policies are outside the scope of this report.	Loan guarantee; lending support mechanisms; grant.
Indonesia 3rd	**Installed Capacity** 1,002 MW **Resource Potential** 27,510 MW[c] **Percentage of Power Grid Capacity** 3%[d]	Indonesia's state-owned utility, PT PLN, has a monopoly on electricity distribution and pricing (DOS 2012). Geothermal development is decentralized, with the main developers being state-owned enterprises. However, public-private joint ventures and private companies are increasingly undertaking development activities.	Government-led exploration.

Table 1. (Continued)

Country and Installed Capacity Rank	Geothermal Information	Background	Exploration Support Type
New Zealand 6th	**Installed Capacity** 731 MW **Resource Potential** Additional 1,100 MW[e] **Percentage of Power Grid Capacity** 7.5%[f]	Due in part to the abundant resource, geothermal power is a significant source of electricity in New Zealand. The country generated 13.7% of its power from geothermal plants in 2012 (Ministry of Economic Development 2013). Although two of the five power producers in New Zealand are now private firms, most of the country's geothermal exploration occurred as the result of government-led efforts (GeothermEx 2010).	Government-led exploration.
Iceland 7th	**Installed Capacity** 670 MW **Resource Potential** 4,300 MW[g] **Percentage of Power Grid Capacity** 22%	Iceland has a competitive electricity market, although the state-owned firm Landsvirkjun dominates wholesale electricity production with an almost three-quarter market share. Nearly 100% of Iceland's power comes from renewable sources, with about one-quarter of that deriving from geothermal production (Orkustofnun 2011).	Grant; government-led exploration.
Japan 8th	**Installed Capacity** 535 MW **Resource Potential** 23,500 MW[h]	Japan has a deregulated and liberalized electricity sector. Ten regional utilities are responsible for the country's electricity supply and cooperate via the Federation of Electric Power Companies of Japan to promote a stable and efficient national power system. In 2012, the country was powered by 90% fossil fuels. However, the recently implemented feed-in tariff is	Grant; government-led exploration.

Country and Installed Capacity Rank	Geothermal Information	Background	Exploration Support Type
	Percentage of Power Grid Capacity ~0.2%[i]	expected to spur the renewable energy market (DLA Piper 2012).	
France 18th	**Installed Capacity** 15 MW in Guadeloupe; no electricity production on mainland **Resource Potential** Low for hydrothermal Percentage of Power Grid Capacity Less than 0.2%[j]	France has a liberalized and deregulated electricity sector that is the second-largest in the European Union (behind Germany). Although the country does not have significant conventional geothermal potential in its mainland territory, it does feature the oldest drilling insurance scheme in Europe (developed in 1982). Despite the lack of conventional geothermal resources for power generation on mainland France, the country hosts a prominent enhanced geothermal system research and development facility in Soultz-sous-Forets that includes an operating power plant.	Drilling failure insurance; grant.
Germany 22nd	**Installed Capacity** 7.3 MW **Resource Potential** Low for conventional hydrothermal Percentage of Power Grid Capacity Less than 0.1%[j]	The German power market is the largest in Europe (Maroo 2012). The country is a worldwide leader in renewable deployment, generating nearly 20% of its electricity from renewable resources, though geothermal power represents only a marginal fraction of this percentage (GEA 2012a). The German government is increasing its support for geothermal development—both in domestic and foreign markets—with lending programs through its development bank, Kreditanstalt für Wiederaufbau Bankengruppe (KfW), and premium feed-in tariff rates.	Loan guarantee; lending support mechanism.

Table 1. (Continued)

Country and Installed Capacity Rank	Geothermal Information	Background	Exploration Support Type
Australia 25th	**Installed Capacity** 0.1 MW **Resource Potential** Low for hydrothermal[k] **Percentage of Power Grid Capacity** Less than 0.01%[l]	Most of the electricity in Australia is delivered through two large electricity markets and two vertically integrated utilities. Currently, the country has minimal geothermal power on its grid with an installed capacity of 0.1 MW. However, in 2009 Australia offered a grant designed specifically to cover a portion of the cost of exploration drilling. Due to a lack of hydrothermal resources, this subsidy is suited toward the development of enhanced geothermal systems, as demonstrated in the Geodynamics-Gooper Basin project discussed with Bendall et al.[m] However, the design of the grant itself could apply to the full range of geothermal energy uses, including conventional hydrothermal. The grant is no longer available.	Grant.

All installed capacity figures are from GEA (2012a) except figures for the United States, which are from GEA (2013); France, Germany, and Australia, which are from BNEF 2012; and New Zealand, which is from Ministry of Economic Development (2012). Figures for geothermal potentials and percent of total capacity derive from: [a]Williams et al. 2008; [b]EIA 2012; [c]GEA 2012a; [d]Darma et al. 2010; [e]Harvey et al. 2010; [f]Ministry of Economic Development 2012; [g]Íslandsbanki 2010; [h]Akeno and Sugino 2010; [i]GEA 2012a; [j]EIA 2013; [k]GeothermEX 2010; [l]Beardsmore and Hill 2010;, and [m]Bendall et al. 2013.

In addition to these older programs, the federal government has enacted or revised a number of policies and incentives since the mid-2000s that provide support to geothermal projects, both in general and potentially to the exploration phase more specifically. These programs were:

- The Section 1703 Loan Program enacted in 2005 and its extension, the Section 1705 Loan Program enacted in 2009
- The Bureau of Land Management's *Programmatic Environmental Impact Statement for Geothermal Leasing in the Western United States*, intended to help streamline the federal leasing system and define a set of best practices for hydrothermal development in the western United States (Doris et al. 2009)
- The ITC, PTC, and changes to these programs including the 1603 Treasury Grant (Speer 2012)
- Support for innovative early-stage drilling technologies through the Recovery Act and DOE Funding Opportunity Announcement #0000522.

4. Policy Descriptions

Below, we define five generic policy types that are commonly used to spur renewable energy deployment in general and, in some cases, geothermal development in particular. Importantly, we have categorized these policy types by their relative leverage or the degree to which public expenditures can attract private investment.

For example, in the case of a loan guarantee program, the government does not actually set aside capital equal to the amount of the private loans that are being backstopped. It merely has to set aside enough capital to cover the probability of defaults in the entire portfolio of guaranteed projects (Mendelsohn 2010). In contrast, when the government assumes the full responsibility of drilling wells and does not incentivize the private sector to do so, the government is accountable for the full cost contributed, not including any exploration activities undertaken by private sector developers (i.e., the exploration could be a joint effort). It is important to note that actual leverage will depend on the specifics of individual policy design.

Finding a balance between leverage and subsidy levels is important because if a subsidy is too high and does not leverage private funds, as in the case of government-led exploration, public funds may be overcommitted and expended rapidly. On the other hand, if a subsidy is too low, very few investors may take advantage of the policy because it may not sufficiently reduce exploration risks.

Due to the emphasis in the United States on market-driven policies with minimal government expenditures, we assessed policies primarily by their ability to leverage private investment. There are several additional ways to assess the effectiveness of a policy, such as total cost, installed capacity, market growth, increase in capital flows, and barrier mitigation. While considering the following analysis of each policy type, it may be helpful to keep in mind that leverage is not the exclusive indicator of policy success.

Table 2 orders the five main policy types analyzed by leverage; the rankings are purely conceptual and do not account for any modifications in policy design or measurements of actual impact.

4.1. Loan Guarantees

Loan guarantees are one type of credit enhancement that can strengthen a borrower's ability to take on debt by improving the project's risk profile. Loan guarantees involve three parties: the borrower, the lender, and the guarantor. In the case of a government policy, the guarantor is usually a government agency or public financial institution. In the event the borrower defaults on debt payments, the guarantor pays off the remaining loan to the lender on behalf of the borrower, usually up to some agreed-upon limit (e.g., 80% of the total loan value). Depending on the policy structure, the guarantor may obtain ownership of project collateral in a default situation as a means of offsetting the cost of the loan repayment.

Loan guarantees and loan forgiveness are similar policies with some distinctions between them from a policy perspective, but could both be considered. From the perspective of the borrower, loan forgiveness is similar to a loan guarantee. Under a loan forgiveness program, a government provided loan could either be repayable no matter the outcome of exploration drilling, or partially or fully forgivable in the case of unsuccessful drilling. This can take the form of a loan being transformed into a grant (in which case, no payment is required). An example can be found in Germany's "Risk of Non-Discovery of Deep Geothermal Energy" program, described in detail in Section 5.3. Thus, under either type of program, the borrower is no longer obligated to repay a portion or all of a loan in the event the project is "unsuccessful" as defined by a given program.

Table 2. Illustrative Qualitative Assessment of Leverage Capability by Policy

<table>
<tr><th>Low Leverage</th><th>Medium Leverage</th><th>High Leverage</th></tr>
<tr><td>Government-led exploration: Low; government incurs full cost of exploration and investment forfeiture in the case of dry wells</td><td>Lending support mechanisms: Medium; interest from the loans could help defray costs, provided that the default rate remains low</td><td rowspan="2">Loan guarantee: High in the case of limited guarantee payouts
Drilling failure insurance: High in the case of limited claims</td></tr>
<tr><td colspan="2">Grants and cooperative agreements: Medium to low; represent a liability in either the case of direct payouts or foregone tax income</td></tr>
</table>

[a] The assessments of leverage provided here are general comparisons across the five policy types. Actual leverage will depend on the specifics of policy design.

As best as possible, loan guarantee programs need to accurately assess the risk of default to ensure loan defaults can be covered and also to accurately communicate the potential cost of the program to the public.

4.2. Drilling Failure Insurance

Drilling failure insurance—also referred to as geologic risk insurance or exploration risk insurance—transfers risk from one party (the originator of the risk) to another (the insurer) in exchange for payment of a premium that reflects the probability of a loss occurring because of that risk. In the case of geothermal drilling, the developer collects a payout if exploration drilling is unsuccessful per a predetermined definition of success. By shifting some of its risk exposure away from its balance sheet, the original party has preemptively managed, in theory, a portion of its potential losses. Financiers that are confident their investments are partially backstopped by creditworthy entities (e.g., a reputable private insurer or national government) may be more likely to furnish capital (and perhaps on more favorable terms) than would be possible without the insurance.[6]

A drilling failure insurance program can be a potentially high-leverage policy option, provided the insured pool of projects have a success rate that is as high or higher than anticipated (and thus revenue from premiums collected cover all claims). Similar to loan guarantees, administrators of drilling failure insurance programs also need to plan for the expensive due diligence process to assess the project risk; however, this could be covered as part of the premium.

4.3. Lending Support Mechanisms

Lending support mechanisms, or “soft” loans, directly reduce the cost of financing as compared to loan guarantees and drilling failure insurance, which indirectly reduce the cost of capital. Like loan guarantees, lending support mechanisms involve a third party, which can be a government agency, public financing institution, or a development bank. The third party provides some form of a capital transfer to the lender for reducing the borrower’s financing costs. As discussed in Section 3.1, lending support mechanisms may not provide as much leverage for private investment as loan guarantees or drilling risk insurance; funds are generally paid out regardless of the loan’s performance and thus are unrecoverable funds. However, lending support mechanisms could be an effective way to spur market development and thus may leverage a large amount of private investment. There is the potential with lending support mechanisms that financiers may have a conflicting incentive to charge higher rates on top of the government subsidy. Checks must be in place to ensure the borrower is actually obtaining a lower financing cost.

There are multiple ways to subsidize both public and private loans, including the following:

- Private loans
  - *Interest rate buy-downs* in which a public entity pays a portion of the interest rate required by the private lender
- Government-provided loans
  - *Lower interest rates*, which can be provided via a direct loan from a public entity
  - *Interest-free periods* during which no interest accrues on the outstanding capital
  - *Payment-free periods* in which the borrower makes no payments to the lender, although interest may continue to accrue
  - *Longer loan terms* than what could be found in the private market, thereby reducing payment levels (although the cumulative interest cost could be higher with a longer amortization period)

- *Repayment bonuses* or small grants that are received as a reward for timely repayment of a loan.

4.4. Grants and Cooperative Agreements

Grant and cooperative agreement programs (grants), unlike drilling failure insurance or loan guarantees, immediately reduce the private investment required for the high-risk exploration phase. Furthermore, grants are less risky to the project developer than tax incentives. The developer receives the value of the incentive right away, which is not dependent on the developer's future, unknown income levels in order to monetize the incentive. Generally, grants can take the form of direct grants of a set amount or a cost-share that covers a predetermined percentage of total exploration costs.

Another type of grant is a *repayable grant.* A repayable grant is similar to loan forgiveness and is repaid to the government entity in the event a project meets a predetermined level of success. By recouping public funds when a well is successful, this method can help increase the leverage and sustainability of the program. However, it may provide a disincentive for fully developing a project if it is on the borderline of being successful and depending on the level of the grant. Thus, the grant must be high enough to incentivize risk taking without being so high that it is more lucrative to take the grant than to develop a project.

Some grants may come with stipulations or conditions. Matching-funds requirements oblige the developer to raise a predetermined ratio of outside investment, leveraging the public outlay. In another example, programs may require developers to provide data about the hydrothermal resources encountered during the exploration process for the dual purpose of: (1) lowering the risk of future exploration and (2) improving industry processes and technologies. Stipulations and conditions can help government agencies address goals beyond simply drilling more wells. But these same requirements can also make the exploration process more onerous and expensive for the project developer.

4.5. Government-Led Exploration

A government ministry, agency, or government-run utility may contract work to private industry to identify and explore a geothermal resource.

Likewise, a government-run utility could undertake this work directly. Subsequently, the government entity may either:

- Issue a tender for a private developer to build and operate a power plant at the site of a proven resource or do so itself
- Execute land lease agreements with developers if an identified but unproven resource is on public lands.

With a government-led exploration program, the government may pay only a portion or all of the exploration expenses.

The goals of direct government participation in geothermal exploration may vary. One goal could be to reduce barriers to private investment by reducing or eliminating the risks of the initial resource identification and exploration phases and may be applied when the risks are perceived to be too high to be addressed by other policies. Another goal could be that of the government- run utility wanting to add to and diversify the energy mix, address unmet power demands, or operate a profitable enterprise.

5. Exploration Policy Case Studies

The following section presents select case studies of the policies that are the focus of this report. These case studies were selected based on their applicability to the U.S. market and are organized by country rather than by policy type as some countries have implemented multiple policies. Each sub-section begins with a brief summary of the geothermal market in the particular country to provide contextual detail and then discusses the primary details of the policies involved. A number of these examples do not correspond directly to the stylized policies presented above. Some illustrate the different ways that the five main policy types have been combined or adapted to specific national contexts. In Section 6, we discuss the policy examples together and the applicability of the various policy types to the United States.

5.1. United States: Loan Guarantees, Lending Support Mechanisms, and Grants and Cooperative Agreements

The United States has an installed capacity that exceeds that of any other country by more than 1 GW. Most existing hydrothermal power plants in the

United States are concentrated in California and Nevada, but significant resources remain untapped in these and other states. The U.S. federal government and some states have attempted to support additional capacity additions with a variety of policies designed to incentivize private sector development beyond what might occur solely due to market forces.

The policies described below are those that have directly or indirectly provided support to the exploration phase of the project development cycle; a fuller list is included in Section 3. Though the United States has implemented several other measures to support the geothermal market (see Section 3), the focus of this analysis is policies specifically applicable to the exploration phase.[7] For example, from 2009 to 2011, projects were eligible for the 1603 Treasury Cash Grant but could only receive the grant after becoming operational.[8]

The list of policies included within this report is not intended to be comprehensive, but rather it highlights a few key programs that are most relevant to the scope of the report, which focuses on policies to encourage the private sector's exploration of geothermal resources. For example, we do not include a discussion of the Program Research and Development Announcement because it focused on feasibility studies only (Lund et al. 2012).

We also do not include analysis of the state-coupled case studies program as it provided support to states and universities, whereas the focus of this report is public-private partnerships.[9]

The United States has provided two loan guarantee programs: one direct loan program, and one cost-share grant both of which could have been applicable to the exploration phase of hydrothermal geothermal projects. Additionally, the federal government has supported exploration for geothermal heat and combined heat and power projects. We have included these programs because the support mechanism may be applicable to supporting exploration activities more broadly, regardless of the intended end-use. We have omitted a discussion of the Section 1703 Loan Program in this section because it has not supported any geothermal projects to date,[10] and because we do explore the Section 1705 program, which is similar to the 1703 program in its design.

Table 3 summarizes the programs that are reviewed in greater detail within this section.

Table 3. U.S. Loan Guarantees, Lending Support Mechanism, and Grants and Cooperative Agreements

Year	Policy Name	Notes
Loan Guarantees		
1974	Loan "Guaranty" Program	The original geothermal loan guarantee program was enacted with passage of the *Geothermal Energy Research, Development and Demonstration Act* in an attempt to catalyze private lending to geothermal projects.
2009	Section 1705 Loan Program	This program was enacted by the Recovery Act as an addition to the 1703 Loan Program. The 1705 program was originally funded with $6 billion, but this amount was later reduced to $2.5 billion and included some modifications compared to the 1703 program.
Lending Support Mechanisms		
1980	Loans for Geothermal Reservoir Confirmation Program	Authorized by Congress, the program was intended to provide exploration loans to both geothermal power and heat projects. Power projects could borrow a maximum of 50% of project costs, but they were not to exceed $3 million. Although authorized, the program never received congressional appropriations.
Grants and Cooperative Agreements		
1977	Program Opportunity Notices (PONs)	This program offered a cost-share to geothermal heat projects for exploration drilling and the demonstration of geothermal energy uses, including combined heat and power.
1978	Industry-Coupled Case Studies Program	This cost-share mechanism covered 20% to 50% of the exploration and reservoir confirmation costs. In exchange, developers had to provide drilling and well data.
1980	User-Coupled Confirmation Drilling Program	Along the same lines as the Industry-Coupled program, the User-Coupled program provided cost-share grants to conduct exploration drilling for geothermal heat projects. The government's portion of the expenses ranged from 20% to 90%, depending on the degree of success achieved in the drilling.

5.1.1. 1974 Geothermal Loan Guaranty Program

In the 1970s, geothermal projects struggled to access commercial loans because lenders were reluctant to invest in an unfamiliar technology (Owens

2002). The original geothermal loan guaranty program was enacted in 1974 with passage of the *Geothermal Energy Research, Development, and Demonstration Act* (Bloomquist et al. 2005; DOE GTP 2012b).[11] It was created to address this reluctance among private industry investors by having the federal government share the risk of early-stage geothermal project development (Owens 2002). The program, which came into effect June 25, 1976, was the first loan guarantee in the United States to support the development of energy resources (Nasr 1978). Specifically, it was established to underwrite loans for the:

1. Determination and evaluation of the resource base
2. Research and development with respect to extraction and utilization technologies
3. Acquisition of rights in geothermal resources
4. Development, construction, and operation of facilities for the demonstration or commercial production of energy using geothermal resources
5. Construction and operation of new commercial power plants.

Step 1, the "determination and evaluation of the resource base," could have included exploration efforts, and others have suggested that supporting exploration activity was in fact one of the objectives of the program (Nasr 1978). However, no exploration projects were supported under this program.

In 1978, Congress amended the 1974 *Geothermal Energy Research, Development and Demonstration Act* with the passage of the *Technical Amendments to the Geothermal Energy Research, Development, and Demonstration Act*. The amendments included raising the maximum guarantee amount to $100 million per project (for a total of $200 million per borrower) and pledging the "full faith and credit of the United States" for payment of the guaranties (Nasr 1978).[12]

Projects were given priority consideration for loan guarantees if they:

- Held the possibility of rapid energy production from geothermal resources
- Demonstrated or used new technologies
- Demonstrated or exploited new geothermal resource areas with commercial potential.

Although the program was designed to support exploration activities, projects that proposed geological or geophysical exploration or the acquisition of land or leases received lower priority when being considered for loan guarantees (Nasr 1978). More information on why this occurred could not be found, although it could have been due to the perceived risks and the related longer project development times. Additional prioritizations were applied that could have impacted the ability of exploration projects to access guarantees (Nasr 1978).

Under the program, the maximum loan term was the lesser of 30 years or "the expected average useful life of any major physical asset to be financed by such [a] loan" (Nasr 1978). Originally, the program guaranteed 100% of loans for up to 75% of projects costs, with the applicant contributing the remaining 25% of capital as equity (Nasr 1978). The program was amended in 1980 to guarantee up to 90% of costs (Bloomquist et al. 2007). This decrease in the equity requirement was likely intended to address the issue—raised by industry during the U.S. House of Representatives' Sub-committee on Energy Research, Development, and Demonstration meetings on the geothermal loan guaranty program—that the 25% equity requirement had been unappealing (Nasr 1978).

The coverage ratio for the program was originally 1:4 and was later reduced to 1:7 by the appropriations committee (Nasr 1978).[13] We could not determine why the appropriations committee altered the coverage ratio. It is possible the committee chose to increase the leverage of public to private dollars to support additional projects.

In addition to the minimum equity requirement, developers supplied project data, and lenders submitted financial reports as part of the guarantee application. Relevant project data included environmental reports; geological, geophysical, and geochemical data; well data; the utilization process; financial information about the firm; project economics; management capabilities of the firm; a milestone plan describing the project goals; marketability of the resources; legal data concerning the project's assets, leases, pending litigation, patents, and permits; and the structure of the organization (Nasr 1978). The loan guarantee application included a description of the management experience of project staff; audited financial statements; the loan agreement; and the lender's assessment of the borrower's loan application. Developers could have included multiple site locations in an application (Nasr 1978).

Table 4. Geothermal Project Defaults on Original U.S. Loan Guaranty Program

Project	Date of Loan Guarantee	Date of Default	Amount	Project Capacity	Type
Westmorland Development Project	1979	1984	$29 million	55 MW	Exploration and field development/ electrical
CU-1 Venture	1980	1984	$49.4 million	Unknown	Exploration/ electric
Niland Development Project	1984	1985	$99.6 million	49 MW	Electrical

Sources: Bloomquist et al. 2007; DOE IG 1987; GAO 1980.

As of February 1978, three loan guaranties had been issued and six applications were under review (Nasr 1978). An estimated 38 additional loans were being prepared with another 12 potential applicants (Nasr 1978). At the conclusion of the program, eight guaranties had been granted (DOE IG 1987). Thus, it seems likely that a significant number of applicants were either not granted guarantees or that applications were rescinded by the developers. The program was scheduled to expire in 1984 and was ended due to a lack of congressional appropriations (Bloomquist et al. 2007; GAO 1980).[14]

Although the original loan guarantee program did support some geothermal development, several projects defaulted on loans, thereby requiring the federal government to make the loans whole (Bloomquist et al. 2007). In all, DOE backed $178 million in defaulted loans from three of the eight projects that received loan guaranties. These projects are outlined in Table 4. For a complete list of geothermal projects that received cost-share or loan guarantee support, see Bloomquist et al. 2007.

According to Nasr (1978), the original Geothermal Loan Guaranty program had a minimal effect on spurring geothermal development, the result being that response to the program in its first four years was much less than anticipated. Nasr argues that other policy changes, such as the streamlining of environmental regulations and the lease application and granting process, as well as the granting of tax write-offs and depletion allowances would have had a greater impact on geothermal development. Nasr (1978) also claimed that the guarantee program did not encourage any new projects that would have otherwise not happened without the guaranty—in other words, the loan guaranty program did not result in additional projects. Other policy changes occurred in the early 1970s, such as implementation of regulations for the federal leasing program, which were not completed until December 1973. The first lease sale was made in January 1974.

According to Bloomquist et al. (2007), the two main limitations for the loan guarantee program were the:

1. Severe eligibility requirements that resulted in loan guarantees being granted to projects that would likely have qualified for conventional loans
2. Reluctance by utilities to participate in the program due to concern about the impact of default on their credit ratings.

Nasr (1978) also noted several specific barriers to an effective loan guarantee program for geothermal exploration projects:

- **Aversion to supporting risky, exploratory projects**: Although the original program was established to support exploratory, high-risk projects that could otherwise not gain financing, the program itself considered the exploration projects to be too risky. However, no official regulations or laws stated that non-exploration projects should be given priority.
    - As an example, as of 1978, one exploratory project—the Beryl and Lund project under development by Geothermal Kinetics—applied for a loan guarantee. Project developers sought support for completing exploration work, drilling temperature gradient holes, and drilling a test well at each site. Geothermal Loan Guaranty administrators determined that these exploratory efforts were too risky, but the program did grant a guarantee for another Geothermal Kinetics site—the Brawley site in Imperial Valley, California, to support completion of test wells.
- **Long application timeframes**: The average time for the three loan guaranties that went through the application process as of 1978 took 16 months, nearly 9 months of which was spent under government review. The application process involved a minimum of 26 steps.
- **Lack of formal application review procedure.**
- **Administrative burden:** The agencies tasked with implementing the program may not have had the organizational capacity to handle the

administrative burden of reviewing applications within the timeframes prescribed by the program itself.

- **Variances in application quality often required additional iterations to provide complete applications.**
- **Absence of clear approval authority among the various organizations involved in the review process.**

In addition to avoiding these barriers, loan guarantee programs in general are advised to establish secure and efficient mechanisms for obtaining collateral in the event of a loan default. An audit of the program by the DOE Office of the Inspector General reviewed the collateral procedures for the three defaulted loans and recommended that DOE (1) refer all cases to the U.S. Attorney General to maximize the value of collateral recovered and (2) prepare final plans to expedite recovery of collateral (DOE IG 1987). These recommendations were in response to long delays in recouping collateral from projects in default. The delays resulted in added costs and a decline in the value of the collateral. Accessing collateral quickly and without additional legal wrangling can help a program keep its overall total losses to a minimum.

Under a loan guarantee for geothermal exploration, the ability to recover collateral could be limited because the developer has yet to obtain any significant assets, such as a power plant. Accounting for the limited collateral in the policy design of any loan guarantee targeting the exploration phase is important for determining the overall cost of the program and how the cost may need to be covered via other means, such as a fee paid by the developer.

5.1.2. Section 1705 Loan Program—2009

In 2009, Congress enacted the Section 1705 Loan Program when it passed the Recovery Act. The Section 1705 Loan Program supported commercialized clean energy projects and manufacturers as an addition to the Section 1703 Loan Program passed in 2005. The 1703 program is omitted from this report, as it has not supported geothermal projects in the past and there is no expectation that it will in the future either (Speer 2012).[15] The Section 1703 program is also similar to the Section 1705 program, and thus a description of one is deemed sufficient.

Unlike the 1703 program, the Section 1705 Loan Program applied to conventional geothermal projects. The 1705 program provided for an important programmatic change: it allowed applicants to participate in DOE's Financial Institution Partnership Program (FIPP). Under FIPP, a private industry lender is required to: (1) be the applicant, (2) fulfill the financial and other due diligence considerations, and (3) provide a certification to DOE that the application is accurate and complete (Pillsbury 2009). In other words, FIPP transfers much of the due diligence from DOE to the private industry lender, which has impetus to conduct accurate assessments due to its "skin in the game" with only 80% of the loan guaranteed. The program was originally funded at $6 billion, which was reduced to $2.5 billion after reallocations (Mendelsohn 2010). The maximum guarantee per project was $500 million (Recovery Act 2009).

FIPP was targeted for commercial projects using relatively simple finance structures and those that had credit ratings of "BB" (S&P or Fitch) or "Ba2" (Moody's) (WSGR 2009; Mendelsohn 2010). Projects were limited to 80% debt, and the program would cover 80% of the loan, or in other words, the program would cover up to 64% of the total capital (Martin 2009; Mendelsohn 2010). Fees, including those for the application, facility, and maintenance for the program varied by loan size but roughly equaled 1% of the loan amount (Mendelsohn 2010).

For each guarantee, the federal government set aside a credit subsidy or a percentage of the total amount toward a pool of funds to cover any losses (Jaffe 2009; Mendelsohn 2010). The credit subsidy is estimated to have been between 6% and 10% of each loan amount, and thus it has been calculated that the program could have supported between $25 billion to nearly $42 billion in total loans for all renewable energy technologies (Mendelsohn 2010). The program expired on September 20, 2011.

Table 5. Geothermal Project Recipients of Federal Loan Guarantees

Project (s)	Developer	Lender	Program	Amount[a]	Capacity	Issued
Blue Mountain	Nevada Geothermal Power Co.	John Hancock	1705, FIPP	$99 million - closed	36 MW	9/2010
Neal Hot Spring	U.S. Geothermal	Federal Financing Bank	1705	$97 million - closed	23 MW	2/2011
McGinness, Jersey Valley, Tuscarora	Ormat Nevada	John Hancock	1705, FIPP	$350 million-conditional	121 MW	6/2011
RETRACTED: Wister, CD-4, Dead Horse Wells[b]	Ormat Nevada	John Hancock	1705, FIPP	$330 million	80 to 90 MW	N/A

Sources: Brightenergy.org 2010; GEA 2010b; Scharfenberger 2011; Ormat 2010a; Ormat 2010b.

As shown in Table 5, four geothermal projects were awarded loan guarantees under the 1705 program; however, one developer retracted its application before it was complete. Two of the three projects that went through with the loan guarantee used FIPP, including the three bundled Ormat plants: Jersey Valley, McGinness, and Tuscarora. In all, the 1705 program could support 180 MW of installed capacity and $545 million in loans for geothermal projects pending project completion.

Of the five plants supported by the three guarantees that proceeded, three plants (Blue Mountain, Tuscarora, and Jersey Valley) are in operation and two projects (Neal Hot Springs and McGinness) are nearing completion (Neubauer 2012; GEA 2012; U.S. Geothermal 2012). It should be noted that both the Blue Mountain and Jersey Valley projects were already in development when granted the loan guarantee (Neubauer 2012; GEA 2012), and thus the guarantees did not support exploration efforts for those projects. Whether the loan guarantees supported or encouraged exploration for the other projects is unclear. However, the 1705 program was capable of facilitating low-cost financing (e.g., a 4.14% interest rate for one project granted a guarantee) (Nevada Geothermal Power 2011).

DOE awarded $545 million in loan guarantees to geothermal projects, while solar photovoltaic and concentrating solar projects received $13.5 billion under the 1705 program (Speer 2012). Possible reasons for the difference in amounts awarded include perceived high transactions costs by developers, disinterest among lenders to participate in the program, lack of geothermal and DOE program expertise on the part of lenders, and a perception of higher risk due to the pursuit of a loan guarantee (Speer 2012). Another possible reason for the limited use of the 1705 program may have been the mismatch between the extended geothermal development process and the short timeframe for participation in the 1705 program (Speer 2012). In light of the timing mismatch of the 1705 program included in the Recovery Act, geothermal project developers and lenders may be able to better use future loan guarantees if they are available over a longer period of time. Also, projects need to be of a certain size to make economic sense given the added cost of going through the loan guarantee application process, which may deter smaller projects.

5.1.3. Loans for Geothermal Reservoir Confirmation Program

Congress authorized the Loans for Geothermal Reservoir Confirmation Program, a direct loan program, in 1980. It was intended to provide loans to geothermal heat and power projects for surface exploration and drilling (Bloomquist 2005; Bloomquist 2007). Loans could not exceed $3 million and were limited to 50% of project costs for power projects. Despite being passed in the Energy Security Act of 1980 and authorized by the Secretary of Energy, the program never received congressional appropriations.

5.1.4. Program Opportunity Notices

The PONs program offered by DOE in 1977 and 1978 was a cost-share exploration drilling grant intended primarily for geothermal heat projects (Bloomquist 2005; Bloomquist 2007; Lund 2012; Moore et al. 2010). Combined heat and power projects were also eligible. DOE received 22 applications for the program and issued at least one PON for a geothermal power project, which was never completed (Moore et al. 2010). Interestingly, the one PON issued specifically for a power project included a provision permitting DOE to recover up to 50% of the grant from the revenues generated.

5.1.5. Industry-Coupled Case Studies Program

DOE implemented the Industry-Coupled Case Studies Program in 1978. The program had two main objectives. First, its cost-share mechanism helped facilitate geothermal exploration by offsetting some of the high initial costs and risk of exploration drilling. At the time, little was known about the locations of geothermal resources in the United States (Moore et al. 2010). To encourage exploration, DOE covered 20% to 50% of the exploration and reservoir confirmation costs for participating projects (Bloomquist 2005). Second, DOE wanted to gather more data on geothermal resources and projects for the purpose of increasing knowledge and thereby aiding in future geothermal power development (Moore et al. 2010). Thus, DOE required industry participants to provide the well and drilling data collected during the exploration process.

During this program, industry partners evaluated 14 sites, eight of which resulted in geothermal power plants. As of 2010, seven of these plants were operating with a combined capacity of 137 MW (Moore et al. 2010); it is possible more plants have come online since then. The program eventually lost congressional support and is no longer in operation.

5.1.6. User-Coupled Confirmation Drilling Program

The User-Coupled and the Industry-Coupled programs served similar purposes but were offered for different end-uses. Through the User-Coupled program, DOE offered a cost-share grant to geothermal heat projects for a range of resource confirmation activities, including drilling (Bloomquist 2005;

Bloomquist 2007; Lund 2012). The program offered to cover 90% of project costs in the case of completely unsuccessful drilling and 20% in the case of completely successful drilling (presumably based on the flow-rate and temperature of the geothermal fluid) (Bloomquist 2005). Specific information on how many grants were awarded and their value was not found.

5.2. France: Drilling Failure Insurance and Grant

The geothermal market in France is geographically fragmented. All of France's conventional geothermal electric capacity (about 15 MW) is located in Guadeloupe, one of its Caribbean departments. In continental France, the Paris Basin and the Aquitaine Basin are low-energy reservoirs that generally do not produce sufficient heat or flow-rates to drive electricity- generating turbines. However, the country does exploit low-energy reservoirs for direct use and drilling for EGS is likely to increase in the near future. There is also a prominent EGS research and development facility in Soultz-sous-Forets that includes an operating power plant. Although the policies described below generally support exploration drilling for direct use or EGS, the mechanisms themselves could apply to hydrothermal drilling as well.

Table 6 summarizes the two programs that are reviewed in greater detail within this section.

Table 6. French Grant and Drilling Failure Insurance

Year	Policy Name	Notes
Grant		
N/A	Grants for Feasibility Studies	This grant program provides up to 50%, but no more than €300,000, of the cost to conduct feasibility studies for heat and power projects. The grant is administered by the French Environment and Energy Management Agency.
Drilling Failure Insurance		
1982	Geothermal Risk Guarantee System	The exploration-phase component of this insurance program provides investors coverage for up to 90% of the cost to drill the first test well in return for a premium payment equivalent to 3.5% of the covered cost in the Île de France region and 5% in all other regions. The program insures the present value of the project once it is in operation. This component does not mitigate exploration risk.

The French Environment and Energy Management Agency (ADEME) administers a grant program to subsidize geothermal energy feasibility studies. This mechanism is somewhat unique in that it does not provide funds for exploration drilling. Instead, project developers—either public or private—can apply for a grant to cover 50% of the cost of a feasibility study for deep geothermal resources, not to exceed €300,000 (GEOFAR 2009a). The available literature does not specify exactly what constitutes a feasibility study according to ADEME; however, the resource identification phase that occurs before exploration drilling generally includes a literature review of existing information and data; geologic surveys leading to a geologic map of the area; geophysical and geochemical surveys; and in some cases, the commencement of temperature-gradient hole drilling (GeothermEx 2010; GEA 2010a).

A consortium of public and private entities also administers the Geothermal Risk Guarantee System, which was the first public insurance program in Europe to address risk in exploration drilling specifically for geothermal projects (GEOFAR 2009a). It has involved two programs: the first one was created in response to the first oil crisis and was in place from 1974 to 1982, and the second has been in place since 1982 (GEOFAR 2009b). These programs have facilitated development of France's two major geothermal resources in the Paris Basin and the Aquitaine Basin, respectively. Although this program supports drilling for power generation projects, all of the wells drilled thus far on the French mainland have been for geothermal heat projects (GEOFAR 2009b).

The current French scheme consists of two mechanisms that have different time horizons and support different stages of geothermal project development. The first mechanism is a short-term insurance program that provides coverage to investors for up to 90% of the cost of the first well if the well does not yield a productive resource. Just as with traditional insurance, this mechanism socializes the risks of geothermal drilling—that is, it allocates the costs of the program among various participants and bundles the proceeds to pay for failed drilling ventures (GEOFAR 2009a). In cases of partial success, the program compensates the participant as necessary to render the project profitable. Compensation is based on an economic model that considers the temperature and flow-rate of the well and the technical, financial, and economic feasibility of the project. The one-time premium to participate in the program is 3.5% of the covered amount in the Île de France region (which includes the productive Paris Basin) and 5% in regions where the risk is higher (GEOFAR 2009b).

The second mechanism is a 20-year insurance contract that guarantees the present value of the well if a resource has indeed been proven. This mechanism compensates a project sponsor in the case of an irreversible decline in the temperature or flow-rate of the well once the project is online and, as such, does not actually support the exploration phase. For this reason we do not provide a more comprehensive review of it.

The Geothermal Risk Guarantee System is managed by SAF-Environment, a partnership of ADEME, Caisse des Dépôts et Consignations (a financial institution under control of the French Parliament), and other private financial entities (GEOFAR 2009b). The program fund is capitalized with payments from participants and project proceeds, investments, and public funds from ADEME (Bezelgues-Courtade et al. 2008).

France currently has another risk guarantee for large groundwater heat pumps (AQUAPAC), but as this technology falls outside the scope of this report, it will not be discussed here.

5.3. Germany: Loan Guarantee and Lending Support Mechanism

Low-energy geothermal resources are being exploited for direct use and combined heat and power applications in Germany, and developers have recently begun to look into deep, hot aquifers and EGS applications (Schellschmidt et al. 2010). The German government and KfW, the national development bank, support the development of geothermal heat and power through a number of policies and financial incentives.

Table 7 summarizes the two programs that are reviewed in greater detail within this section.

Germany offers multiple financial support programs for renewable energy projects under the Market Incentive Programme (MAP) that is administered by the German Federal Ministry for the Environment, Nature Conservation, and Nuclear Safety (BMU) and the German development bank KfW (BMU 2012; Wendel et al. 2010). BMU and KfW offer two programs for geothermal power projects: the Risk of Non-Discovery of Deep Geothermal Energy program and the Renewable Energies Programme for Deep Geothermal Energy. For the purposes of this analysis, we do not assess the latter because it does not apply to the exploration phase.

In 2009, Germany launched the Risk of Non-Discovery of Deep Geothermal Energy program (Fündigkeitsrisiko Tiefengeothermie) to minimize exploration risks associated with geothermal projects. The unique feature of this program is that it combines exploration risk mitigation with project financing (Kreuter and Schrage 2010). For the purposes of this report, we have categorized this program as a hybrid loan guarantee and a lending support mechanism because KfW, a government entity, is the lender and offers 100% loan forgiveness in the case of unsuccessful drilling.

BMU and KfW provide this unique financing mechanism via a €60 million revolving loan fund. The maximum loan size for a given project is the lesser of €16 million or 80% of the eligible drilling costs (KfW), with the developer covering the remaining 20%. The loan period can be as long as 10 years. During the drilling period, KfW requires a risk surcharge to be paid in the form of a higher interest rate, usually in the range of 10% to 20%. Additionally, a €65,000 application fee is required along with an additional €45,000 for an approved loan to cover the program's monitoring and auditing costs.

There are two important features of the loan: a one- to two-year grace period during which the developer can defer repayment and a full release from loan liability mechanism. The full release from loan liability is enacted if the developer does not achieve a predetermined success threshold (e.g., if the thermal capacity of a drilled well is less than the required level to make the project viable) (Kreuter and Schrage 2010). If the exemption is enacted, the developer will not be required to repay the loan, and KfW will cover up to 80% of the drilling costs with the developer covering the remaining portion.

Table 7. German Loan Guarantee and Lending Support Mechanism

Year	Policy Name	Notes
Loan Guarantee		
2009	Risk of Non-Discovery of Deep Geothermal Energy	The loan program includes a loan forgiveness provision of up to 100% of the money borrowed by the developer in the case of failed exploration drilling. We have classified this provision as a loan guarantee.
Lending Support Mechanism		
2009	Risk of Non-Discovery of Deep Geothermal Energy	The program offers loans up to €16 million or 80% of the eligible drilling costs for developers to undertake exploration drilling. Because KfW, the lender, is the national development bank of Germany, we consider this program a government-provided loan and have classified it as a lending support mechanism in addition to a loan guarantee.

An important feature of the Risk of Non-Discovery of Deep Geothermal Energy program is that developers are required to apply for the program through the local commercial bank with which they have a pre-existing relationship. For their part, the commercial bank processes the application and guarantees the loan to KfW in the event the developer goes bankrupt (though this guarantee does not cover drilling failure) (Kreuter and Schrage 2010). The developer's bank also facilitates development of the loan application, which is then submitted to KfW (Kreuter and Schrage 2010). By requiring the bank to guarantee the loan, the program induces the bank to thoroughly assess the risks. Municipalities, municipally owned enterprises, and municipal special-purpose associations can file their loan applications directly to KfW.

The unique feature of this program, as compared with U.S. loan guarantee programs, is that a government entity provides the financing that is guaranteed by a private lender. The U.S. loan guarantee programs, in comparison, represent guarantees provided by the government for debt supplied by private sources. Another unique feature of the German program is that KfW attempts to account for possible losses due to the high risk of geothermal exploration by charging a higher interest rate, called a risk premium, during the drilling phase. Under this feature, the additional interest charged protects public resources from excessive losses; however, the risk premium may prove to be too expensive to increase the level of project activity. Thus, the Risk of Non-Discovery of Deep Geothermal Energy program serves as an example of how one government created a customized approach to supporting its geothermal market by developing an innovative policy that incorporates elements of loans, loan guarantees, and insurance.

Although no data is publically available regarding the number and amounts of loans made under KfW's Risk of Non-Discovery of Deep Geothermal Energy Program, geothermal development in Germany is progressing. As of July 2012, four combined heat and power plants were operating with a total installed capacity of about 7.9 MWe, and another 30 projects were in development (BNEF 2012).

5.4. Multilateral Banks and Development Agencies: Drilling Failure Insurance and Grant

Superficially, the following three programs—Geofund, ARGeo, and the Geothermal Risk Mitigation Facility for East Africa—implemented by multilateral banks and development agencies may not appear to be relevant to the U.S. policy and market context. Geofund is now defunct, while ARGeo has faced serious delays.[16] All three were or are administered and funded through multilateral agencies or development banks. The United States, as a developed country, does not receive funding or other support from multilateral institutions for the development of domestic energy infrastructure, nor does it have a national development bank as do some other developed countries (e.g., Germany). However, the United States could still replicate these risk mitigation mechanisms summarized in Table 8.

5.4.1. Geofund

In 2006, the World Bank launched the GeoFund, a program to facilitate geothermal development in eastern Europe and central Asia (Shimizaki 2008).[17] The fund was seeded with a $25 million grant from the Global Environmental Facility but was dropped in 2009 after spending its first tranche of about $4.5 million (World Bank).

Table 8. Multilateral Banks and Development Agencies Drilling Failure Insurance and Grants

Year	Policy Name	Notes
Drilling Failure Insurance		
2006	GeoFund	Established by the World Bank to support geothermal energy development in eastern Europe and central Asia, GeoFund's main mechanism was drilling failure insurance. The World Bank discontinued the program after the pilot program in Hungary failed to confirm sufficient flow-rates and collected a USD $3.3 million claim.
2009	ARGeo	The Global Environment Facility established ARGeo in partnership with other multilateral institutions in order to provide technical assistance and exploration risk mitigation to geothermal projects in the East African Rift region. Originally structured as a drilling failure insurance scheme, the risk mitigation mechanism was eliminated when the World Bank rescinded its funding.
Grant		
2012	Geothermal Risk Mitigation Facility	The German development bank KfW, in partnership with the European Union and African Union, offers grants for geothermal exploration and drilling. The program is targeted to projects in five nations located along the East African Rift. The fund pays additional success fees once the project secures outside funding for subsequent phases of development.

The GeoFund program was divided into three component instruments: technical assistance, direct investment funding for projects, and a drilling failure insurance mechanism. The only project that received coverage under the insurance mechanism was a pilot project in Hungary sponsored by the Hungarian Oil and Gas Company (MOL) and geothermal developers Enex (Iceland) and Vulkan Kft (Australia). The project did not proceed to construction primarily because sufficient flow-rates were not discovered in the two test wells. As a result, the drilling failure insurance mechanism was activated, and the World Bank paid MOL $3.3 million out of a coverage limit of $3.72 million (the limit was set at 85% of the eligible drilling cost) (Shimizaki 2008; World Bank 2010).

5.4.2. ARGeo

In partnership with the World Bank and United Nations Environment Programme, the Global Environment Facility established the East African Rift Geothermal Development Program (ARGeo) in late 2009 to provide technical assistance and exploration risk mitigation funding (Mwangi 2010). ARGeo is intended to support operations in six target countries to exploit resources along the seismically active East African Rift that are largely undeveloped.[18]

The original ARGeo proposal took a dual approach: (1) regional networking facilitation and technical assistance and (2) a drilling risk mitigation fund that exclusively addressed exploration- drilling risk. In exchange for a premium paid by the developer, the risk mitigation fund provided up to 85% of eligible drilling expenses in the event that exploration drilling yielded an insufficient resource. This coverage ratio would be reduced for each successive well attempted. The World Bank rescinded its funding for the risk mitigation portion of the project in December 2011, but at the time of writing, the networking and technical assistance portion of the project was still active.

5.4.3. Geothermal Risk Mitigation Facility for Eastern Africa

After pulling out of the planning process for ARGeo, KfW created a geothermal risk mitigation facility in partnership with the European Union and the African Union in late 2011.[19] This program offsets the risk of non-performing exploration wells by providing grants for 80% of the cost of surface studies, 40% of the cost of exploration drilling, and 20% of the cost of required infrastructure improvements (GRMF 2012). Grants are available to both government and private developers. To further incentivize developers, the bank offers an additional financing success fee for projects that secure outside financing for subsequent phases of development (KfW 2011; Muir 2011). Although it offers additional incentives and funding for other activities in addition to exploration drilling, this facility essentially functions in much the same way as the U.S. Industry-Coupled Case Studies program: it is a cost-share program with a data collection requirement.

At the time of publishing, the Geothermal Risk Mitigation Facility had just closed the first round of applications. KfW and the African Union have stated their intentions to expand the facility after it is piloted.

5.5. Australia: Grant

Conventional hydrothermal resources are nearly non-existent in Australia. However, the country has significant potential for developing deep hot aquifers and EGS resources, and the mechanisms Australia is using to support the development of these resources could also be applied to conventional hydrothermal exploration.

Table 9 summarizes the program that is reviewed in greater detail within this section.

In 2008, Australia introduced the Geothermal Drilling Program to provide $50 million Australian dollars (AUD) for exploration drilling. The government sought to offset a portion of the exploration drilling expenses for a limited number of projects by offering matching grants of up to AUD $7 million. The Australian Department of Resources, Energy, and Tourism issued seven grants through a competitive, merit-based selection process (Australia DRET 2009).

Of the seven grants awarded, four were forfeited. In a report commissioned by the Australian Centre for Renewable Energy, the Allen Consulting Group (2011) noted that the developers found it difficult to raise the required matching funds by the set deadline (due in part to unfortunate coincidence with the global financial crisis); this may have been a primary factor for the forfeitures. The final round of applications closed in August 2009, and all funded projects must be completed by December 31, 2014.

5.6. Iceland: Grant and Government-Led Exploration

Geothermal plants provide a significant portion of Iceland's electricity needs and represent nearly one-quarter of installed capacity and generation. Due to the country's location on the mid- Atlantic ridge, the country has advantageous conventional hydrothermal resources. Another unique aspect of geothermal development in Iceland has been its decentralized and yet public nature. Iceland's government exploration activities date to 1969 when a municipal government constructed the first geothermal power plant in the country. Since then, resource exploration, exploration drilling, development, and electricity production have remained mostly the responsibility of municipal, regional, or national government entities, under the regulation of the National Energy Authority (GeothermEx 2010).[20]

Table 10 summarizes the two programs that are reviewed in greater detail within this section.

Iceland launched the National Energy Fund in 1967 to provide loans and grants to support geothermal exploration drilling (heat and electricity applications). In 2003, Iceland passed an act that placed administration of the program with the National Energy Authority (Björnsson 2010). The fund includes a few programs for various types of geothermal projects, but hydrothermal exploration is eligible for the fund's grant program only (Government of Iceland 2009). Public and private organizations are eligible for the grant that can cover up to 50% of project costs, making it a cost-share program (GEOFAR 2009c). The grant is paid out in three equal payments: (1) when work begins, (2) when work is halfway done, and (3) when work is complete (GEOFAR 2009c).

Table 9. Australian Grant

Year	Policy Name	Notes
Grant		
2008	Geothermal Drilling Program	Australia implemented a cost-share grant program to support exploration drilling for EGS and deep, hot aquifers. The program provided a matching grant of up to AUD $7 million for drilling activities.

Table 10. Icelandic Grant and Government-Led Exploration

Year	Policy Name	Notes
Grant		
2008	National Energy Fund Grant	The National Energy Fund offers a cost-share grant to support exploration activities up to 50% of project costs. The grant is paid out in three equal payments.

Table 10. (Continued)

Year	Policy Name	Notes
Government-Led Exploration		
1969	Government-led Exploration	Geothermal energy development has been led by government entities at both the national and municipal levels. Currently, most exploration still takes place as the result of municipal government and/or utility geothermal development. There is only one private sector entity that operates power plants.

5.7. Japan: Grant and Government-Led Exploration

Japan was an early adopter of geothermal energy for power generation. The first commercial plant was commissioned in 1966 (Akeno and Sugino 2010). At over 23 GW, Japan has one of the highest potential hydrothermal resource bases in the world. The power market is restructured and dominated by 10 regional electric utility companies. Geothermal capacity additions have been stagnant since the mid-1990s, but recently adopted renewable energy policies and incentives may be positioning the geothermal industry for a revival.

Table 11 summarizes the two programs that are reviewed in greater detail within this section.

In 1974, the Japanese Agency of Industrial Science and Technology (AIST) undertook a large geothermal research program that included exploration drilling (GeothermEx 2010). In 1980, the government transferred responsibility for the program to the New Energy Development Organization (NEDO). Together, the two agencies funded the research initiative for the purpose of reducing exploration risk and subsequently contracted private firms to undertake the drilling (GeothermEx 2010). As the research initiative identified and demonstrated resource effectiveness, NEDO and AIST transferred their use to private firms to develop them into power plants (GeothermEx, 2010).[21]

In conjunction with its research program, the Japanese government introduced a cost-sharing program to subsidize exploration drilling. Although not well funded in recent years, the program covers 50% of the cost to drill exploratory wells, with 100% repayment required if the wells are successful (Akeno and Sugino 2010). In 1986, the government expanded the program to also cover 20% of the cost of development and injection wells (Akeno and Sugino 2010).

The reduction in risk created by NEDO's cost-share subsidy and research effort may have contributed to geothermal power development in Japan. Installed capacity increased from 139 MW in 1976 to 539 MW in 2011 (BNEF 2012; GeothermEx 2010). Although the cost-share program still existed as of 2010, Japan's geothermal power sector stagnated between 1999 and 2010, adding just 2 MW in 2004 (Akeno and Sugino 2010).

5.8. New Zealand: Government-Led Exploration

New Zealand was an early adopter of geothermal energy. Commissioned in 1958, the Wairakei geothermal power plant was one of the first in the world. New Zealand currently gets a significant portion of its power from geothermal plants, which generated 13.7% of the country's electricity in 2012 (Ministry of Economic Development 2013). This is due in part to the market competitiveness of geothermal power and other renewables in New Zealand, and also to the abundance of hydrothermal resources on the North Island (Harvey et al. 2010).

Table 12 summarizes the program that is reviewed in greater detail within this section.

Table 11. Japanese Grant and Government-Led Exploration

Year	Policy Name	Notes
Grant		
2008	Cost-Share Grant	The cost-share grant covers 50% of the cost to drill exploratory wells and 20% for development and injection wells. In recent years it has not been well funded.
Government-Led Exploration		
1974	Geothermal Research Program	The Agency of Industrial Science and Technology (AIST) created a geothermal research program in 1974 in response to the first oil crisis. The program included exploration drilling and was transferred to the New Energy Development Organization in 1980.

Table 12. New Zealand Government-Led Exploration

Year	Policy Name	Notes
Government-Led Exploration		
1958	Government-led Exploration	Most of New Zealand's geothermal development occurred when the country's utility companies were government-owned. As such, these government utility companies, and in turn the government itself, carried the risk associated with exploration drilling.

Most of New Zealand's geothermal power development was due to government-led exploration (GeothermEx 2010). Although two of the five electricity producers in New Zealand are now private sector companies, much of the country's geothermal development took place when the government bore the exploration risk, and some of the government-run utilities still undertake geothermal exploration projects today (GeothermEx 2010).

5.9. Indonesia: Lending Support Mechanism and Government-Led Exploration

Indonesia has the highest potential for geothermal power in the world, but with only 1,002 MW of installed capacity, much of the potential remains untapped (BNEF 2012). PT PLN, the state- owned utility, has a monopoly on electricity distribution and pricing. Traditionally, state-owned enterprises and municipal entities conducted geothermal development activities, but public-private partnerships and private developers have started to enter the industry.

Table 13 summarizes the two programs that are reviewed in greater detail within this section.

In 2011, Indonesia created the Geothermal Fund Facility (GFF), a revolving loan fund with an initial capitalization of 1.237 trillion Indonesian rupiah ($145 million USD) designed to help mitigate exploration-drilling risk.[22] The GFF is structured as a public-private collaboration in which local governments or firms (private and state-owned) in possession of either a geothermal mining permit or a geothermal power concession undertake exploration activities (Hasan and Wahjosoedibjo 2012; PIP). The GFF issues loans to these three types of entities to undertake exploration activities.

When a local government receives funding from the GFF, they contract a private firm to undertake the exploration activities, including drilling

(Wahjosoedibjo 2012). Once a potential resource proves viable for power generation, the local government can issue a tender for development of the proven resource. Then, the winning bidder must pay for the exploration data in order to receive its mining license. Proceeds from the data and information package purchased by the winning bidder will partially recapitalize the fund (Hasan and Wahjosoedibjo 2012; PIP). After the winning bidder develops a field and builds a power plant, the bidder sells its power to the government-owned utility through a power purchase agreement. The Indonesia Investment Agency will likely forgive loans to local governments for any unsuccessful exploration activities (Wahjosoedibjo 2012).

With private or state-owned firms that are holders of geothermal mining permits and geothermal power concessions, GFF funding takes the form of a loan disbursed in stages that is repayable with interest when the entity completes successful or unsuccessful exploration activities (PIP). Because the loan is repayable whether or not the resource proves viable, the entity conducting exploration activities bears all risk of unsuccessful drilling. Private developers and state-owned enterprises that successfully develop steam fields into operating power plants sell their power to the state-owned utility through power purchase agreements.

Table 13. Indonesian Lending Support Mechanism and Government-Led Exploration

Year	Policy Name	Notes
Lending Support Mechanism		
2011	Geothermal Fund Facility	This government loan program provides private sector developers, state-owned enterprises, and municipal government developers with loans for exploration and confirmation from a revolving loan fund. Municipal government entities can access the funding without repayment in the case of failed exploration drilling, while private sector and state-owned enterprise developers must repay the loan with interest, no matter the outcome of exploration drilling.
Government-Led Exploration		
1974	Geothermal Research Program	Until recently most geothermal development had been conducted by government entities, such as state-owned enterprises and/or government-owned utilities. More recently, private sector developers and public-private partnerships have begun to enter the market; however, most existing geothermal plants were the result of government-led efforts.

Because both government entities (such as municipal governments and state-owned enterprises) and private developers develop geothermal resources in Indonesia, the country's policy regime can be categorized as a hybrid lending support mechanism and government-led exploration. The GFF is a government-provided lending program available to private developers, government entities, and public-private partnerships. The state-owned developers and municipal governments that develop geothermal resources give the hybrid regime its government-led character. In the case of local government exploration, the GFF takes on the risk of failed exploration drilling. On the other hand, private and state-owned developers assume all the exploration risk under the GFF. However, this type of government exploration scheme is different from government agencies undertaking exploration drilling and power plant construction directly because the government-owned utility neither develops nor operates the steam field or power plant.

6. Characteristics of Exploration Policies and Applicability to the United States

This section examines which overall policies and/or specific policy components spurred geothermal deployment, along with providing possible explanations as to why they were successful. Each discussion includes an assessment of the appropriateness of a given program to the U.S. policy context. Due to a lack of detailed information beyond what is available publically and the subsequent challenge of linking impacts to a particular policy, this analysis is predominantly qualitative. As in Section 4, the following discussion is organized by potential leverage of private investment (highest to lowest).

6.1. Loan Guarantee

Loan guarantee programs are generally public-private partnerships and involve multiple contracts between several parties. Thus, these programs tend to have complex structures, and their effectiveness can be impacted by many different elements of the policy's design.

It is important for the guarantor to accurately assess the ability of geothermal exploration projects to access loans on the private market. The guarantor must also determine whether providing a loan guarantee will bring down the cost of financing by an amount that will help projects move forward. In addition, these programs have the potential for high losses in the case of loan defaults, although due diligence in assessing project and borrower risks could help mitigate this possibility.

The U.S. 1974 Loan Guaranty and Section 1705 programs described in Section 5.1 would seem to be an obvious fit to be reapplied in the United States. However, for a couple reasons, these programs may not be good candidates for implementation today. First, there may be less political appetite for another loan guarantee program following the high profile bankruptcies and defaults of a few borrowers under the 1705 program. Second, due to fiscal concerns, it may be difficult to gain enthusiasm for a program in which the ultimate cost is unknown. But the creation of a loan loss reserve could make a loan guarantee more appealing. Once the loan loss reserve is expended, any subsequent defaults would be made whole through private channels (i.e., there would be no government backing after a certain amount of defaults).

With regard to a loan forgiveness-type guarantee, such as Germany's Risk of Non-Discovery of Deep Geothermal Energy program, this type of guarantee is less likely to apply to the United States than the traditional loan guarantees previously offered by the U.S. federal government. The United States has no national development bank akin to KfW to administer such an initiative, and the federal government usually prefers to take a more market-oriented approach by incenting private lenders with government loan guarantees rather than providing loans. Furthermore, the KfW program has found it challenging to engage developers' commercial banks because many lack experience assessing the risk of geothermal projects, thereby creating high per-loan transaction costs (Kreuter and Schrage 2010). Programs looking to replicate this model may want to consider how best to engage private-sector lenders in geothermal financing, especially where there has been little experience to date. For example, banks could be provided with training on geothermal technologies and project risks.

However, DOE or another government agency, such as the Federal Financing Bank, could be authorized to provide loans directly with a forgiveness mechanism in much the same way that the Department of Education provides loans for higher education. In fact, the Federal Financing Bank has provided a direct loan to a geothermal project–guaranteed by the Section 1705 Program—in the past. The Neal Hot Spring project recently put into commercial operation by U.S. Geothermal received a non-forgivable loan of nearly $100 million from the Federal Financing Bank (Lowder 2012).

Despite this success, most U.S. government agencies do not have the capacity to manage loans directly as KfW does in Germany.

Interestingly, low-cost public-private financing programs have been implemented at the state level, and there were several attempts in Congress and the Senate to establish a government- backed finance institution for clean energy projects in 2009.[23, 24] The bills that were introduced died in committee. Generally, using the model of direct loans would convey more risk and less leverage of private investment to public resources as compared to providing loan guarantees.

Another attractive feature of the Risk of Non-Discovery of Deep Geothermal Energy program is that it seeks to support alternative geothermal power technologies, such as enhanced geothermal systems, and it does not require a typically requisite probability of success (POS) study (Kreuter and Schrage 2010). By implementing a similar program in the United States, there is the potential to encourage projects that use technologies or applications beyond those that are currently being commercially developed, such as EGS.

6.2. Drilling Failure Insurance

As demonstrated in the case of the Geofund (Section 5.4), public insurance programs have high leverage potential on a programmatic basis (although claims paid to a given project in the event of an unsuccessful exploration can be high). Provided that these programs can judiciously select clients and diversify risk exposures, drilling failure insurance programs can be a cost-effective and self-sustaining option.

The French example shows how a risk insurance scheme could enable development at scale while containing losses. However, this is due in part to the high success rate of drilling for geothermal heat projects. The available data on the wells drilled in the Paris Basin since 1987 (when France reinstituted its insurance scheme) show a success rate of 74% (with partial failures at 17% and total failures at 9%), which is substantially higher than the rate encountered by geothermal power developers in the United States (Bezelgues-Courtade et al. 2008). In the United States, developers can expect a success rate somewhere in the neighborhood of 35% to 50% for well-researched and vetted drilling sites (Speer 2012). If the United States were to implement a similar program, it could potentially offset the lower success rates by charging a higher insurance premium.

Effective risk management entails the optimal balance of returns to risks. For drilling insurance schemes to be self-sustaining, the revenue stream from premium payments must be higher than paid claims. However, premium rates should not be so high as to discourage developer participation.

In addition to the applicability considerations gleaned from the case studies in Section 5, policymakers may want to consider how to provide clarity in the claims-making and payout processes. Clear information about the required capital requirements, policy holder eligibility, the division of payouts between the developer and investors, eligible claims, and the coverage triggers (i.e., at what temperature and flow-rate is the well considered a loss) can help lower the participants' perceptions of policy risk. Specifying how many test wells are eligible for cost recovery and whether there is a declining rate of coverage for each well can also help prevent developers and program administrators from making erroneous assumptions about covered losses.

Government insurance programs currently exist in the United States primarily to allow risk- prone populations without access to the private insurance market the opportunity to protect themselves against certain exposures. Such gaps in private coverage usually exist because either the cost of insurance is out of their reach for those that need it most or insurance for particular risks is not available. The earthquake insurance offered by California is one example. Catastrophes, such as earthquakes and floods, pose potentially devastating and concentrated loss probabilities, and private insurers may risk insolvency if their coverage in these areas is triggered at large scale. Therefore, the federal and state governments have stepped in to meet a need (i.e., the protection of its population) that is in their interest.

If geothermal energy deployment is deemed to be in the national interest (e.g., for the various environmental and energy security benefits it provides), it may be reasonable for policymakers to devise a scheme that mitigates or removes one of its principal barriers: the exploration drilling risk.

An insurance program would likely require some high up-front costs. It would need to staff an underwriting team as well as an administrative body, and the cost of conducting due diligence on the initial projects would not be defrayed by the proceeds from an existing portfolio. Such a program could sustain itself over time but would likely require several years to do so, depending on the rate of project participation.

6.3. Lending Support Mechanisms

Lending support mechanisms for the exploration phase of geothermal development may be an attractive policy alternative due to the fact that nearly all exploration-phase financing in the United States comes from equity investors. Not only is equity generally more costly than debt, but the lack of lenders willing to make loans for exploration drilling and reservoir confirmation significantly limits developers' potential sources of capital at this phase. Lending support mechanisms might spur more lenders to fund geothermal projects during the exploration phase, rather than solely at the production/injection well drilling and power plant construction phases.

In the U.S. context, a lending support mechanism in the form of an interest rate subsidy could attract private lenders by providing interest payments sufficient to cover the risk of exploration drilling, while at the same time providing developers with a lower cost of capital. In contrast, government-provided loans could crowd out private investment. With direct government loans, private capital might instead flow to subsequent phases of development. To have a significant impact on project development, such a program would require substantial public funding with the possibility for losses. This puts into question the applicability of a government-provided loan program in the United States, at least at the federal level. This is especially true when considering recent concerns surrounding defaults under the DOE loan guarantee program and general concerns about fiscal prudence.

The structure of Indonesia's GFF provides an innovative example of a government direct loan program. Should U.S. policy seek to offset exploration-drilling risk faced by project developers through some sort of government loan program, a revolving loan fund may provide the most value to taxpayers when compared to other types of government loan programs, grants, and cooperative agreements. Because it is likely that 90% of U.S. geothermal resources lie on federal land (DOI 2008), the government could tender proven resources for leasing, as Indonesia does. The initial public investment in exploration drilling could then be repaid by the winning bidder from the capital they raise from private investors for subsequent phases of development. In the United States, such a fund could be managed as a national geothermal trust fund, which is a familiar concept for several other federally funded programs.

6.4. Grants and Cooperative Agreements

Grant programs directed specifically at exploration drilling reduce the amount of equity and/or debt a developer must raise from outside sources prior to the exploration phase.

The U.S. Industry-Coupled Case Studies Program, Iceland's National Energy Fund grant program, and the Japanese cost-share program all may have helped contribute to the early development of the geothermal power sector within those countries. As previously mentioned, the Industry-Coupled program in the United States resulted in the exploration of 14 sites and the construction of eight power plants, seven of which remained in operation in 2010 (Moore et al. 2010). Likewise, the Japanese cost-share program contributed to a significant increase in installed capacity in the late 1970s and early to mid-1990s. Iceland's grant program, in concert with municipal and regional government-led efforts, was a key driver of the growth of the geothermal power sector. Today, at least 22% of Iceland's total installed generation capacity is geothermal (Ragnarsson 2010). These three examples demonstrate the potential usefulness of grants and cooperative agreements adapted specifically for the exploration phase of geothermal development.

A grant or cost-share covering a portion of the exploration drilling and resource confirmation costs helps developers to mobilize outside finance before having proven the resource. Furthermore, programs, such as Japan's cost-share program, may increase the sustainability of public funding by requiring repayment of funds from successful projects or requiring winning bidders to repay the exploration costs on proven resources.

The country programs described in Section 5 highlight the fact that the impact of such programs on individual projects can depend on per-project funding levels, the structure of payments, and government capacity for funding them. An overly ambitious deadline for securing funds to match the AUD 7 million grant in the case of Australia's Geothermal Drilling Program may have impeded developers' ability to use the grant (Allen Consulting Group 2011). As a result, developers had to forfeit more than half of the grants awarded. On the other hand, the U.S. Industry-Coupled program generously covered up to 50% of exploration drilling costs (Bloomquist 2003).

Additionally, grants covering exploration activities other than drilling, such as France's feasibility study grant, may do little to address the significant exploration risk associated with drilling. But these studies may decrease the proportion of failed exploration wells as they can refine potential resource areas and nominally reduce the cost of developing projects.

TEXT BOX 1. DRILLING TAX INCENTIVES

Geothermal power projects cannot currently take advantage of renewable energy tax incentives (e.g., the PTC and ITC) until a project is operational. Because of the long lag time between project initiation and commissioning, this creates a timing mismatch between the time when the project encounters high exploration risk and when an investor can make use of tax incentives. This mismatch hinders the ability of current U.S. renewable energy tax incentives from addressing the high risk of exploration drilling.

While not grants, tax incentives can create additional value by reducing a developer's tax liability. U.S. oil and gas investors enjoy a number of tax incentives applicable to the exploration and development phase of oil and gas projects. Among those is the option to expense or capitalize intangible drilling costs (IDCs), such as drilling time, labor, and drilling fluids. Investors in oil and gas projects can deduct up to 100% of a project's IDCs in the first year regardless of whether or not a well produces oil (Sherlock 2011). With the passage of the Energy Tax Act of 1978, Congress permitted geothermal developers to enjoy the same tax treatment. However, the business practices of applying this deduction, as well as the percentage depletion deduction, are uncertain and may warrant further research.

Those interested in creating a new capital grant program in the United States that specifically targets exploration drilling expenses may take note of Japan's cost-share program as an alternative to unrecoverable grants. As previously mentioned, Japan attempts to recover all or a portion of the government capital from successful projects in the form of a loan, while waiving repayment for unsuccessful wells. This creates a more sustainable model and higher leverage of private investment. In cases where drilling is successful, repayment of the funding with interest lessens the downside for potentially drilling dry holes. This type of "grant" could also mobilize additional private investment in geothermal power projects by limiting their exposure to only a portion of the exploration risk, in the case of a cost-share. Indeed, a study of risk mitigation strategies funded by the Geothermal Technologies Program in 2008 recommended a government co-funded drilling mechanism to provide capital for exploration drilling (Deloitte Development LLC 2008).

One potential downside is that grants are unrecoverable expenditures, and thus, unless they are "repayable" grants, the government will not recoup the funds once spent. Although the Japanese loan/grant mechanism attempts to recoup a portion of the public investment, the high failure rate of exploration drilling does not permit any grant structure to recoup capital from every funded project.

6.5. Government-Led Exploration

Government-led exploration programs can spur project development but require significant funding and have high administration costs. Also, government-led exploration may not provide much leverage in the exploration phase, as compared to other policy options. However, these types of programs may provide leverage in terms of private sector investments in later phases of development.

The research and demonstration initiatives undertaken by NEDO and AIST in Japan contributed to the early growth of the Japanese geothermal power sector. The efforts of both agencies also highlight the utility of transferring resources proven through government exploration drilling— either through a public tender, land lease, or another mechanism—to the private sector.

Likewise, Iceland's government-led development, in conjunction with the previously described grant, has resulted in substantial industry growth and project development. The unique characteristic of Iceland's government-led development is that municipal and regional governments and publically owned utilities, rather than national government entities, have undertaken nearly all of the development activities. Similar efforts could be a feasible for U.S. municipal utilities and cooperatives located in areas with good, undeveloped hydrothermal resources (or perhaps where EGS is feasible, once technology costs have come down). However, given the geographic location of most hydrothermal resources in the United States, this option is very limited in its overall potential to rapidly develop the country's resources. And with a limited rate base and requirements to make conservative investments, these typically smaller, local utilities may not find geothermal exploration economically viable.[25]

Another complicating factor for government-led development is that governments face the same risks in identifying and exploring geothermal resources as the private sector. Any expenditures that result in dry wells are sunk costs at the expense of the taxpayer. Altogether, most policy supports for the energy sector in the United States, renewable or otherwise, consist of tax incentives and/or subsidies. Unveiling a government geothermal exploration initiative would be inconsistent with recent renewable energy incentives and may not be feasible, especially in light of the fact that the U.S. geothermal market is viewed as more mature than it was in the 1970s and 1980s when the first exploration programs were implemented.

We did not include an analysis of the characteristics or applicability of Kenya's state-owned Geothermal Development Corporation or the Philippines National Oil Company Energy Development Corporation because we do not see such state-owned geothermal development entities as a viable alternative in the U.S. market. For a brief discussion of the Geothermal Development Company and Philippines state-owned energy company, see Appendices A and B.

7. KEY FINDINGS AND ADDITIONAL ENABLING OPPORTUNITIES

The United States has significant, untapped hydrothermal resources. One of the main hurdles to utilizing this baseload electricity resource is the high risk of exploration drilling. Because of this risk, developers find raising capital for this phase of development particularly challenging. Even when developers are able to locate willing investors, the cost of capital can be prohibitive and is almost exclusively equity capital, which is by its very nature more expensive than debt. To address this barrier in attracting investment to the exploration phase, several countries including the United States have implemented policies intended to reduce the cost of capital or mitigate the investment risks. Our analysis has attempted to identify the policies used and comment on their characteristics and potential applicability to the United States.

The five stylized policy types initially presented in Section 4 include loan guarantees, drilling failure insurance, lending support mechanisms, grants and cooperative agreements, and government-led exploration. These policies have been implemented in a number of countries. At various times since the 1970s, the United States has implemented loan guarantees, a lending support mechanism, and grant programs in relatively straightforward manners. On the other hand, applications of all five policy types have been more nuanced when applied in other countries. In some cases, the policies were implemented in a hybridized manner, combining features of more than one policy. In Japan, for example, the government has provided loans for geothermal exploration that convert to grants if a viable resource cannot be obtained. The government-owned German development bank KfW offers a loan program for geothermal exploration in Germany that is notable for its loan-forgiveness feature that functions essentially like a loan guarantee.

Also, countries such as Iceland and the United States have provided multiple policies concurrently. In Iceland, government- or municipally owned companies undertake most of the geothermal exploration projects while the national government provides grants to support the projects. In the United States, the Industry-Coupled Case Studies program operated concurrently with the 1974 Loan Guaranty program and other incentives.

One noticeable difference between the U.S. policies and those of other countries is that the United States has not provided financing or lending support mechanisms designed specifically for the exploration phase. Although the Loans for Geothermal Reservoir Confirmation program was intended to fill this role, it was never funded. Japan, Germany, Indonesia, multilateral banks and development agencies, on the other hand, have provided loans and/or loan guarantees specifically targeted to the exploration phase. Thus far, U.S. loan guarantees have helped facilitate loans for subsequent phases of development or for the entire project cycle. This may be an area for further consideration, especially given the fact that current U.S. policies support equity investments for operational projects.

There may be room, even when considering the current U.S. policy context, for innovative policies and incentives targeted specifically to the exploration phase. The following options are theoretical and would deserve further scrutiny and research to understand their potential efficacy:

- Encourage exploration-phase financing
  - Exploration-phase loan guarantees—Current and past loan guarantees offered by the federal government covered loans for the entire project cycle. Targeting geothermal loan guarantees to the exploration phase may encourage lenders to begin loaning money for exploration (likely on a portfolio basis), a phase currently funded almost exclusively by high cost equity.
  - Direct loan program—Although the 1980 Loans for Geothermal Reservoir Confirmation Program was never funded, authorizing and funding a new loan program specifically targeted to conventional geothermal exploration, or renewable energy

projects in general, may help address the lack of debt financing currently observed in the U.S. market. Loans could either be repayable no matter the outcome of exploration drilling, or partially or fully forgivable in the case of unsuccessful drilling.. The Federal Financing Bank has already provided one loan to a geothermal project. Creating a more targeted and/or dedicated loan program, such as the Green Bank proposed in 2009, could fill in the gap left by private lenders during the exploration phase.
  - Offer a cost-share grant program for resource identification and exploration activities along the lines of the Industry-Couple Case Studies program.
- Exploration-phase tax incentives
  - Tax deduction or credit applicable to all or a subset of exploration costs—The incentives could be offered until projects reach a specific point in the development cycle. Considering the GEA's four-phase development cycle, all drilling costs up to and including the drilling of one full-sized discovery well could be considered. To avoid abuse, the incentive could be paid out in installments corresponding to defined milestones during the project development cycle. The incentives could take the form of a refundable tax credit or cash grant akin to the 1603 Treasury program.
- Expand the resource base by leveraging oil and gas drilling through co-production
  - Incentives for the assessment and/or development of geothermal resources produced as a result of oil and gas drilling—These incentives could include those mentioned above regarding exploration-phase tax incentives, or some other co- production-specific incentive to develop or collect data about co-produced geothermal fluids.

An area for future research could be to quantify the outcomes of the policy applications presented in Section 5 in order to systematically identify and analyze how many projects these policies supported in their respective countries. Such research could also determine how much public funding was required to support projects, as measured against a country's existing resource and market barriers. Except for the U.S. experience, the current analysis did not collect data at this level of detail from each country. Such research could attempt to determine, either through quantitative methods or case-study comparisons, whether policies that provide the most leverage of private investment facilitate the most rapid project development and in which contexts this may be the case. The ultimate purpose of such research would be to identify which policies maximize the goals of deployment and leverage of private investment, either concurrently or exclusively. The current state of slow growth in geothermal development demonstrates that when left to market forces alone, geothermal development will proceed slowly and sporadically.

APPENDIX A. PHILIPPINES: GOVERNMENT-LED EXPLORATION

The Philippines has the second-highest installed geothermal power capacity in the world with 1,998.8 MW, and its 2009–2030 National Energy Plan lays out a strategy for surpassing the United States as the world leader (BNEF 2012; Philippine DOE 2009). The GEA identified the country as having 3,500–5,730 MW of potential. It set a target of 3,447 MW in its 2009–2030 Energy Plan—a 75% increase over the country's 2009 capacity—though it has only installed approximately 27 MW since establishing its most recent target.

The renewable energy sector in the Philippines—the geothermal sector in particular—has recently undergone significant changes. Prior to 2007, government entities had conducted nearly all geothermal development activities in the country, except for the first two steam fields, developed by Chevron's Filipino precursor. In the mid-1970s the government created the Philippine National Oil Company-Energy Development Corporation (PNOC-EDC), which undertook all geothermal development projects until it was privatized in 2007 (Catigig 2008). It is now called the Energy Development Corporation. The World Bank and the Japan Bank for International Cooperation financed the majority of PNOC-EDC's geothermal development activities, providing loans for projects accounting for 1,043 MW of PNOC-EDC's 1,149 MW of installed capacity (Dolor 2005). Due to the fact that the PNOC-EDC was a government-owned company, the national government of the Philippines took all the risk associated with exploration activities.

The privatization of the PNOC-EDC, along with the market-based approach to renewable energy development laid out in the 2008 renewable energy law, will provide an interesting case study of the effectiveness of government-led geothermal development as compared to a market-based approach. However, a full analysis of the transition to a privatized geothermal sector with market-based incentives may not be possible until actual capacity additions coming online since 2007 can be compared to the plan's 2015 target of 2,382 MW.

Over the last decade, the government has opened up the geothermal power sector to competitive market forces. This has included privatization of the state-owned National Power Corporation's geothermal assets in 2001; the introduction of a bidding process for exploration projects in 2008; and the legal definition of geothermal resources as mineral resources, which has allowed entry of foreign development companies into the Filipino market. In 2011, the Philippine government approved contracts for the construction and power purchase of six new geothermal facilities.

Appendix B. Kenya: Government-Led Exploration

The Government of Kenya adopted a more innovative government-led approach to overcoming exploration-phase barriers, which had been stalling geothermal development, by creating the Geothermal Development Company, a wholly state-owned enterprise. To accelerate the previously slow geothermal project development timeframe and to reduce exploration risk, the company undertakes activities including detailed surface well exploration, infrastructure development, and drilling of exploration wells (CIF 2011). After completing exploration drilling, the company intends to tender proven resources to private investors for the subsequent development of power plants. The company's current project at the Menengai geothermal steam field aims to produce 800 MW by 2018 (GDC 2011).

References

Akeno, T. & Sugino, H. (2010). "2010 Country Update for Japan." Proceeding from World Geothermal Congress 2010. Accessed June 25, 2012: http://www.geothermal-energy.org/pdf/IGAstandard/WGC/2010/0142.pdf.

Allen Consulting Group. (2011). "Australia's Geothermal Industry: Pathways for Development." Prepared for the Australian Centre for Renewable

Energy. Accessed July 2, 2012: http://www.allenconsult.com.au/resources/acgpathwaysdevelopment2011.pdf.

American Wind Energy Association (AWEA). (2012). *AWEA U.S. Wind Industry Second Quarter 2012 Market Report*. http://www.awea.org/Resources/Content.aspx?ItemNumber=5364.

Australia Department of Resources, Energy and Tourism (Australia DRET). (2009). "Guide to Managing GDP Funding." Canberra: Government of Australia. Accessed August 10, 2012: http://www.agea.org.au/geothermal-energy/publications-reports/2009/guide-to-managing- gdp-funding/.

Barbut, M. (2012). "African Rift Geothermal Development Facility." Project Document May 31, 2012. Washington, DC: Global Environment Facility. Accessed October 9, 2012: http://www.thegef.org/gef/sites/thegef.org/files/documents/document/05-31-2012%20ID2119%20Council%20letter.pdf.

Beardsmore, G. & Hill, A. (2010). "Australia – Country Update." World Geothermal Energy Congress 2010. Accessed October 25, 2012: http://www.geothermal-energy.org/pdf/IGAstandard/WGC/2010/0150.pdf.

Bendall, B., Huddlestone-Holmes, C. & Goldstein, B. (2013). "Current Status of Geothermal Projects in Australia – A Review." Proceedings from the Thirty-Eighth Workshop on Geothermal Engineering, Stanford, California. Accessed August 8, 2013: https://pangea.stanford.edu/ERE/db/GeoConf/papers/SGW/2013/Bendall.pdf.

Bezelgues-Courtade, S. & Jaudin, F. (2008). "The French Geothermal Risk Guarantee System." Presented at the Workshop on Geothermal Risk Insurance: The World Bank Geothermal Energy Development Program (Geofund). Accessed June 10, 2012: https://pangea.stanford.edu/ERE/pdf/IGAstandard/GeoFund/Germany2008/Bezelgues- Courtade.pdf.

Björnsson, S. ed. (2010). "Geothermal Development and Research in Iceland." Reykjavik: Orkustofnun (National Energy Authority). Accessed June 6, 2013: http://www.nea.is/media/utgafa/GD_loka.pdf.

Bloomberg New Energy Finance (BNEF). (2012). "Global Renewable Electricity Installations Dataset." London: Bloomberg New Energy Finance.

Bloomquist, R., Petty, S. & Wagner, R. (2007). *Geothermal Risk Mitigation Instrument and Incentive Program*. World Bank. Accessed August 13, 2012: http://www.africa- union.org/AURGOKT/docs/english/Geothermal%20Risk%20Mitigation.doc.

Bloomquist, R. (2005). "The Evolution of U.S. Policy Designed to Encourage Geothermal Development Provision of Access and Encouraging Project Development." World Geothermal Congress. Accessed February 25, 2012: http://www.geothermal-energy.org/pdf/IGAstandard/WGC/2005/2201.pdf.

Bloomquist, R. (2003). "United States Geothermal Policy - Provision of Access and Encouraging Project Development." Accessed July 2, 2012: http://www.os.is/gogn/flytja/JHS-Skjol/IGC2003ShortCourse/05Bloomquist.pdf.

BMU (German Federal Ministry for the Environment, Nature Conservation, and Nuclear Safety). (2012). "Guidelines for the Support of Measures to Utilise Renewable Energies in the Heat Market." Berlin: Federal Ministry for the Environment, Nature Conservation, and Nuclear Safety. Accessed October 2, 2012: http://www.erneuerbare- energien.de/english/renewable_energy/downloads/doc/48594.php.

Brightenergy.org. (December 6, 2010). "Ormat Seeks $330m Loan for Three Geothermal Projects." Accessed August 26, 2011: http://www.brighterenergy.org/20399/news/geothermal/ormat-seeks-330m-loan-for-three- geothermal-projects/.

Catigig, D. (2008). "Geothermal Energy Development in the Philippines with the Energy Development Corporation Embarking into Power Generation." Presented at the United Nations University Geothermal Training Programme 30th Anniversary Workshop, 26-27 August 2008. Reykjavík: United Nations University Geothermal Training Programme. Accessed July 18, 2012: http://www.unugtp.is/page/30th-anniversary.

Climate Investment Funds (CIF). (2011). "Scaling-Up Renewable Energy Program (SREP): Investment Plan for Kenya." Accessed July 2, 2012: http://www.climateinvestmentfunds.org/cifnet/sites/default/files/Kenya%20SREP%20Investment%20Plan%20-%20Endorsed.pdf.

Darma, S., Harsoprayitno, S., Setiawan, B., Sukhyar, R., Soedibjo, A., Ganefianto, N. & Stimac, J. (2010). *Geothermal Energy Update Geothermal Energy Development and Utilization in Indonesia.* World Geothermal Congress 2010. Accessed June 10, 2012: http://www.geothermal- energy.org/pdf/IGAstandard/WGC/2010/0128.pdf.

Deloitte Development LLC. (2008). "Department of Energy--Office of Energy Efficiency and Renewable Energy Geothermal Program: Geothermal Risk Mitigation Strategies Report." Department of Energy Geothermal Technologies Program. Accessed February 25, 2012: http://www1.eere.energy.gov/geothermal/pdfs/geothermal_risk_mitigation.pdf.

DLA Piper. (2012). "Japan's Renewable Energy Feed-in Tariff Regime." Accessed November 5, 2012: http://www.dlapiper.com/files/Publication/b01bd22c-1b39-48ca-93f9-124fc9553f2d/Presentation/Publication Attachment/e10d7d94-4ef3-428a-a785-15d88867f2c7/japans-renewable-energy-feed-in-tariff-regime.pdf.

DOE GTP (U.S. Department of Energy - Geothermal Technologies Program). (2012a). "About the Program." Accessed July 8, 2012: http://www1.eere.energy.gov/geothermal/about.html.

DOE GTP. (2012b). "A History of Geothermal in the United States." Accessed June 19, 2012: http://www1.eere.energy.gov/geothermal/history.html.

DOE IG (U.S. Department of Energy - Office of the Inspector General). (1987). "The Department of Energy's Collection Procedures for Defaulted Geothermal Loan Guaranties." Accessed August 10, 2012: http://www.osti.gov/bridge/servlets/purl/6306438- vRlbsV/6306438.pdf.

DOE LGP (U.S. Department of Energy - Loan Guarantee Program). (2012). "FAQs." Accessed July 2, 2012: https://lpo.energy.gov/?page_id=2687.

DOI (U.S. Department of the Interior). (October 2008). "Kempthorne Launches Initiative to Spur Geothermal Energy and Power Generation on Federal Lands." Accessed November 21, 2008: http://www.doi.gov/archive/news/08_News_Releases/102208b.html.

Dolor, F. (2005). "Phases of Geothermal Development in the Philippines." Presented at the Workshop for Decision Makers on Geothermal Projects and Management, November 14-18, 2005. Naivasha, Kenya: United Nations University – Geothermal Development Programme. Accessed September 7, 2012: http://www.os.is/gogn/unu-gtp-sc/UNU-GTP-SC-02-26.pdf.

Doris, E., Kreycik, C. & Young, K. (2009). *Policy Overview and Options for Maximizing the Role of Policy in Geothermal Electricity Development*. NREL/TP-6A2-46653. Golden, CO: National Renewable Energy Laboratory. Accessed July 20, 2012: http://www.nrel.gov/docs/fy10osti/46653.pdf.

DOS (U.S. Department of State). (2012). "U.S. Relations with Indonesia." Accessed June 18 2012: http://www.state.gov/r/pa/ei/bgn/2748.htm.

DRET (Department of Resources, Energy and Tourism). (2009). "Guide to Managing GDP Funding." Canberra: Government of Australia. Accessed August 10, 2012: http://www.agea.org.au/geothermal-energy/publications-reports/2009/guide-to-managing-gdp-funding/.

DSIRE. (2012). "Renewable Electricity Production Tax Credit (PTC)." North Carolina Solar Center. Accessed November 6, 2012.

http://www.dsireusa.org/incentives/incentive.cfm?Incentive_Code=US13F &re=1&ee=1.

EERE (U.S. Department of Energy - Office of Energy Efficiency and Renewable Energy). (2013). "Nevada Deploys First U.S. Commercial, Grid-Connected Enhanced Geothermal System." Accessed May 13, 2013: http://apps1.eere.energy.gov/news/progress_alerts.cfm/pa_id=864.

EERE. (2012). *2011 Renewable Energy Data Book.* Accessed September 4, 2013: http://www.nrel.gov/docs/fy13osti/54909.pdf.

EERE . (2010). *2009 Renewable Energy Databook.* Accessed September 4, 2013: http://www.nrel.gov/docs/fy10osti/48178.pdf.

EIA (United States Energy Information Administration). (2012). "Table 8.11a Electric Net Summer Capacity: Total (All Sectors), 1949-2010." Annual Energy Review. Accessed September 24, 2012: http://www.eia.gov/totalenergy/data/annual/showtext.cfm?t=ptb0811a.

EIA. (2013). "International Energy Statistics." Accessed June 12, 2013: http://www.eia.gov/cfapps/ipdbproject/iedindex3.cfm?tid=2&pid=2&aid=7&cid=AS,&syi d=2009&eyid=2009&unit=MK.

GAO (United States Government Accountability Office). (1980). *The Geothermal Loan Guarantee Program: Need for Improvements.* Prepared by the Comptroller General of the United States for the Chairman, Subcommittee on Committee on Interstate and House of Representatives. Accessed August 10, 2012: http://www.osti.gov/bridge/servlets/purl/5628099- tnYiqW/5628099.pdf.

GDC (Geothermal Development Company). (2011). "GDC Strikes Steam in Menengai." *Steam Magazine.* Nairobi: Geothermal Development Company. Accessed July 2, 2012: http://www.gdc.co.ke/images/Steam/steam4web.pdf.

GEA (Geothermal Energy Association). (2010a). "New Geothermal Terms and Definitions: A Guide to Reporting Resource Development Progress and Results to the Geothermal Energy Association." Accessed February 28, 2013: http://geo-energy.org/pdf/NewGeothermalTermsandDefinitions_January2011.pdf.

GEA. (2010b). "Geothermal Energy Association Statement on Department of Energy Loan to U.S. Geothermal." Accessed August 25, 2011: http://www.geo-energy.org/pdf/press_releases/GEA_DOE_Loan_Statement.pdf.

GEA. (2012a). *Geothermal: International Market Overview Report.* Accessed March 11, 2013: http://www.geo-energy International_Overview.pdf.

GEA. (2012b). *Annual U.S. Geothermal Power Production and Development Report.* Accessed March 11, 2013: http://geo-energy.org/reports/2012 AnnualUSGeothermalPowerProductionandDevelopmentReport_Final.pdf.

GEA. (2013). *Annual U.S. Geothermal Power Production and Development Report.* Accessed March 11, 2013: http://geo- energy.org/pdf/reports/ 2013AnnualUSGeothermalPowerProductionandDevelopmentReport_Final.pdf.

GEOFAR (Geothermal Finance and Awareness in European Regions). (2009a). *Financial Instruments as Support for the Exploitation of Geothermal Energy.* Accessed May 23, 2012: http://www.energia.gr/ geofar/page.asp?p_id=63&cat_id=19&lng=1.

GEOFAR. (2009b). "The Geothermal Risk Guarantee System: France Factsheet." Accessed May 21, 2012: http://www.energia.gr/geofar/ articlefiles/factsheets/France- Factsheet%20InsuranceMechanism.pdf.

GEOFAR. (2009c). National Energy Fund – Iceland. Accessed June 17, 2012: http://www.energia.gr/geofar/articlefiles/factsheets/Iceland%20-%20 Factsheet%20National%20Energy%20Fund%20Loan%20and%20Grant.pdf.

Geotherm EX. (2010). "An Assessment of Geothermal Resource Risks in Indonesia." Accessed July 2, 2012: http://www.ppiaf.org/sites/ppiaf.org/ files/publication/REPORT_Risk_Mitigation_Options_Indonesia.pdf.

Government of Iceland. (2009). "Act on the National Energy Authority in Iceland." Reykjavík: Alþingi (Icelandic Parliament). Accessed August 10, 2012: http://www.nea.is/media/log/NEA- Act_Final.pdf.

GRMF (Geothermal Risk Mitigation Facility for Eastern Africa). (2012). Accessed October 9, 2012: http://www.grmf-eastafrica.org/.

Harbridge House (1977). Federal Loan Guaranties Program Management Report. Accessed June 19, 2012: http://www.osti.gov/bridge/servlets/ purl/7218995-RUH8JC/7218995.pdf.

Harvey, C., White, B., Lawless, J. & Dunstall, M. (2010). "2005 – 2010 New Zealand Country Update." *World Geothermal Congress 2010 Proceedings*, April 25-29, 2010, Bali, Indonesia. Taupo, New Zealand: GNS Science. Accessed November 6, 2012: http://www.geothermal-energy.org/pdf/IGAstandard/WGC/2010/0168.pdf.

Hasan, M. & Wahjosoedibjo, A. (2012). "Geothermal Fund for Hastening the Development of Indonesia's Geothermal Resources." Presented at the Thirty-Seventh Workshop on Geothermal Reservoir Engineering. Accessed June 29, 2012: https://pangea.stanford.edu/ERE/pdf/ IGAstandard/SGW/2012/Wahjosoedibjo.pdf.

Hurlbut, D. (2012). "Geothermal Power and Interconnection: The Economics of Getting to Market." NREL/TP-6A20-54192. Golden, CO: National Renewable Energy Laboratory. Accessed March 25, 2013: http://www.nrel.gov/docs/fy12osti/54192.pdf.

IFC (International Finance Corporation). (2013). "Success of Geothermal Wells: A Global Study." Washington, DC: The World Bank Group. Accessed June 21, 2013: http://www.ifc.org/wps/wcm/connect/topics_ext_content/ifc_external_corporate_site/ifc+sustainability/publications/publications_gpn_geothermal-wells.

Íslandsbanki. (2010). "Iceland Geothermal Energy Market Report." Reykjavík, Iceland: Íslandsbanki. Accessed August 10, 2012: http://skjol.islandsbanki.is/servlet/file/store156/item61173/version3/2010%200419%20Ice land%20Geothermal%20Energy%20Market%20Report.pdf.

Jaffe, S. (4 August 2009). "Robbing Renewable Energy to Pay for Clunkers." Greentech Media. Accessed March 17, 2010: http://www.greentechmedia.com/articles/read/robbing-renewable- energy-to-pay-for-clunkers/.

KfW (KfW Entwicklungsbank). (2011). "More Environmentally-Friendly Electricity From the Ground. East Africa Taps its Geothermal Potential." Press Release, December 15, 2011. Frankfurt, Germany: KfW Entwicklungsbank. Accessed November 6, 2012: http://www.kfw.de/kfw/en/KfW_Group/Press/Latest_News/PressArchiv/2011/20111215_55013.jsp.

KfW. (n.d.). "Renewable Energies." Accessed October 2, 2012: http://www.kfw.de/kfw/en/Domestic_Promotion/Our_offers/Renewable_energy.jsp.

Kreuter, H. & Schrage, C. (2010). "Geothermal Risk Mitigation Schemes in Germany." World Geothermal Congress. Accessed June 19, 2012: http://www.geothermal-energy.org/pdf/IGAstandard/WGC/2010/0404.pdf.

Lowder, T. (2012). "Drilling for Dollars: Notable Developments in Geothermal Finance." Golden, CO: National Renewable Energy Laboratory. Accessed March 11, 2013: https://financere.nrel.gov/finance/content/drilling-dollars-notable-developments- geothermal-finance.

Lund, J. & Bloomquist, R. G. (2012). "Development of Geothermal Policy in the United States – What Works and What Doesn't Work." *Thirty-Seventh Workshop on Geothermal Reservoir Engineering Proceedings*; January 30–February 1, 2012, Stanford, California. Klamath Falls, OR: Geo-Heat Center; 1-10.

Jay, M. (May 2012). "German Power Market Outlook." Risk.net. Accessed June 10, 2012: http://www.risk.net/energy-risk/feature/2172395/german-power-market-outlook.

Martin, K., Greenwald, S., Hansen, K. & Howes, W. (March 2009). DOE Loan Guarantee Webinar. Accessed November 26, 2012: http://www.infocastinc.com/index.php/conference/doeloanweb.

Mendelsohn, M. & Feldman, D. (2013). *The Master Limited Partnerships Parity Act*. Golden, CO: National Renewable Energy Laboratory. Accessed June 3, 2013: http://www.coons.senate.gov/issues/master-limited-partnerships-parity-act.

Mendelsohn, M. (2010). "The DOE Loan Guarantee Program: A Primer." Golden, CO: National Renewable Energy Laboratory. Accessed June 25, 2012: https://financere.nrel.gov/finance/content/doe-loan-guarantee-program-primer.

Meyer, R. T., Coe, B. A., Nasr, L. H. & Bixler, M. (1980). "Geothermal Loan Guaranty Program: Recommendations for Improvement Including New Simplified Application Procedures for Small Projects." Report Abstract. Accessed August 10, 2012: http://www.osti.gov/geothermal/product.biblio.jsp?osti_id=6167295.

Ministry of Economic Development. (2013). "Data Tables for Electricity." Wellington, New Zealand: Ministry of Business, Innovation and Employment. Accessed April 8, 2013: http://www.med.govt.nz/sectors-industries/energy/energy-modelling/data/electricity.

Ministry of Economic Development. (2012). "New Zealand Energy Data File 2012." Wellington, New Zealand: Ministry of Business, Innovation and Employment. Accessed April 8, 2013: http://www.med.govt.nz/sectors-industries/energy/energy-modelling/publications/energy- data-file/new-zealand-energy-data-file-2012.

Moore, J.; Ross, H.; Wright, P. (2010). "Exploration 1976–2006: A History of Geothermal Energy Research and Development in the United States." Washington, DC: U.S. Department of Energy. Accessed July 2, 2012: http://www1.eere.energy.gov/geothermal/pdfs/geothermal_history_1_exploration.pdf.

Muir, J. (2011). "New Fund for Geothermal Power in East Africa." Sydney: Sinclair Knight Merz Consulting. Accessed November 26, 2012: http://www.skmconsulting.com/Knowledge-and-Insights/Achieve-Magazine/Issue3-2011/article5.aspx.

Mwangi, M. N. (2010). "The African Rift Geothermal Facility (ARGeo) – Status." Prepared for UNU-GTP Short Course V on Exploration for

Geothermal Resources, October 29–November 19, 2010. Reykjavík, Iceland: United Nations University Geothermal Training Programme. Accessed November 1, 2012: http://www.os.is/gogn/unu-gtp-sc/UNU-GTP-SC-11-48.pdf.

Nasr, L. (1978). *The Geothermal Loan Guaranty Program and its Impact on Geothermal Exploration*. DOE Contract No. ET-78-C-02-4635. Golden, CO: Colorado School of Mines. Accessed June 21, 2012: http://www.osti.gov/geothermal/servlets/purl/5113359-u5hstI/5113359. pdf.

Neubauer, C. (2012). "Lights Go Dim on Another Energy Project." *The Washington Times*. Accessed August 12, 2012: http://www.washingtontimes.com/news/2012/jul/4/lights-go-dim-on- another-energy-project/?page=all.

Nevada Geothermal Power. (2011). "Blue Mountain: The Power is on at Faulkner 1!" Vancouver: Nevada Geothermal Power, Inc. Accessed October 2, 2012: http://www.nevadageothermal.com/i/pdf/Blue_Mountain_Fact_Sheet_July_2011.pdf.

Orkustofnun (Iceland National Energy Authority). (2011). "Report on Regulation and the Electricity Market." Accessed June 18, 2012: http://www.energy-regulators.eu/portal/page/portal/EER_HOME/EER_PUBLICATIONS/NATIONAL_REPORTS/National%20Reporting%202011/NR_En/C11_NR_Iceland-EN.pdf.

Ormat. (2010a). "Ormat Retracts DOE Loan Guarantee Application for Three GeothermalProjects in California and Nevada." Accessed August 29, 2011: http://www.ormat.com/news/latest-items/ormat-retracts-doe-loan-guarantee-application- three-geothermal-projects-california.

Ormat. (2010b). "Part I of Application for a DOE Loan Guarantee has Been Submitted on July 27th." Accessed August 25, 2011: http://www.ormat.com/news/financing/ormat-names-john- hancock-exclusive-debt-provider-under-doe-loan-guarantee-application.

Owens, B. (2002). *An Economic Valuation of a Geothermal Production Tax Credit*. Golden, CO: National Renewable Energy Laboratory. Accessed August 12, 2012: http://www.nrel.gov/docs/fy02osti/31969.pdf.

Philippines DOE (Philippines Department of Energy). (2009). "Highlights of the 2009-2030 Philippine Energy Plan." Manila, Philippines: Philippine Department of Energy. Accessed July 18, 2012: http://www.doe.gov.ph/PEP/.

Pillsbury. (2009). "The New 'FIPP' Program for DOE Loan Guarantees: Mobilizing Private Capital for Renewable Energy Projects." Accessed

August 11, 2012: http://www.pillsburylaw.com/siteFiles/Publications/0428F04A2E3B865DDA3698E2B1A035D0.pdf.

PIP (Indonesia Investment Agency). (n.d.). "Fund Facility for Geothermal." Jakarta: Pusat Investasi Pemerintah.

Project Finance. (2010). "Hudson Ranch 1: Venture Verified." *Project Finance Magazine*. Ragnarsson, A. (2010). "Geothermal Development in Iceland 2005-2009." Proceedings of the World Geothermal Congress 2010, Bali, Indonesia, April 25-29, 2010. Bochum, Germany: International Geothermal Association. Accessed September 11, 2012: http://www.geothermal-energy.org/304,iga_geothermal_conference_database.html.

Reed, M. (1983). "Assessment of Low-Temperature Geothermal Resources of the United States-1982." U.S. Geological Survey Circular 1982. Accessed July 2, 2012: http://pubs.usgs.gov/circ/1983/0892/report.pdf.

Rickerson, W.; Gifford, J.; Grace, R.; Karlynn, C. (2012). *Geothermal FIT Design: International Experience and U.S. Considerations.* NREL/TP-6A20-53320. Golden, CO: National Renewable Energy Laboratory. Accessed November 6, 2012: http://www.nrel.gov/docs/fy12osti/53320.pdf.

Salmon, J. P., Meurice, J., Wobus, N., Stern, F. & Duaime, M. (2011). *Guidebook to Geothermal Power Finance.* NREL/SR-6A20-49391. Golden, CO: National Renewable Energy Laboratory. Accessed September 4, 2013: http://www.nrel.gov/docs/fy11osti/49391.pdf.

Scharfenberger, P. (2011). "Breaking New Ground: Geothermal Projects Secure Federal Loan Guarantees." Golden, CO: National Renewable Energy Laboratory. Accessed August 25, 2011: http://financere.nrel.gov/finance/content/treasury-cash-grant-transaction-costs.

Schellschmidt, R., Sanner, B., Pester, S. & Schulz, R. (2010). "Geothermal Energy Use in Germany." Proceedings of the World Geothermal Congress 2010, Bali, Indonesia, April 25-29, 2010. International Geothermal Association. Accessed February 21, 2013: http://www.geothermal-energy.org/pdf/IGAstandard/WGC/2010/0152.pdf.

SEIA (Solar Energy Industries Association). (2012). "U.S. Solar Market Insight Q1 2012: Executive Summary." Solar Energy Industries Association and Greentech Media. Accessed August 14, 2012: http://www.seia.org/research-resources/us-solar-market-insight.

Sherlock, M. (2011). "Energy Tax Policy: Historical Perspectives on and Current Status of Energy Tax Expenditures." Washington, DC:

Congressional Research Service. Accessed August 10, 2012: http://www.leahy.senate.gov/imo/media/doc/R41227EnergyLegReport.pdf.

Shimizaki, K. (11 November 2008). "Geothermal Energy Development Program for Europe and Central Asia Region (GeoFund)." The World Bank/GEF. Accessed February 23, 2012: http://www.geothermal-energy.org/pdf/IGAstandard/GeoFund/Germany2008/Shimazaki.pdf.

Speer, B. (2012). "Geothermal Brief: Market and Policy Impacts Update." NREL/TP-6A20-53288. Golden, CO: National Renewable Energy Laboratory. Accessed November 7, 2012: http://www.nrel.gov/docs/fy13osti/53288.pdf.

U.S. Congress. (2009). *American Reinvestment and Recovery Act (ARRA)*. Accessed June 25, 2012: https://lpo.energy.gov/wp-content/uploads/2010/09/ARRA-2009.pdf.

U.S. Congress. (1974). *Geothermal Energy Research, Development, and Demonstration Act of 1974*. Accessed June 19, 2012: www1.eere.energy.gov/geothermal/docs/30-24.doc.

U.S. Geothermal Inc. (2012). "U.S. Geothermal Provides Development Update." U.S. Geothermal Press release. Accessed August 12, 2012: http://www.usgeothermal.com/.

U.S. Geothermal Inc. (n.d.). "San Emidio Project." Accessed August 12, 2012: http://www.usgeothermal.com/SanEmidioProject.aspx.

Wahjosoedibjo, A. (2012). Personal communication, 15 August 2012, PENConsulting.

Watanabe, C. (2012). "Japan Recommends Preferential Price Rates for Renewable Energy." London: Bloomberg New Energy Finance. Accessed June 12, 2013: http://www.bloomberg.com/news/2012-04-25/japan-recommends-preferential-price-rates- for-renewable-energy.html.

Wendel, M., Hiegel, M., Jaudin. F. & Poux, A. (2010). "GEOFAR – Financing Geothermal Energy in European Regions." Accessed February 22, 2012: http://halshs.archives-ouvertes.fr/docs/00/51/48/59/PDF/GEOFARa_Financing-Geotherma-Energy-in-European- Regions.pdf.

Williams, C. F., Reed, M. J., Mariner, R. H., DeAngelo, J. & Galanis, S. P. (2008a). "Assessment of Moderate- and High-temperature Geothermal Resources of the United States." U.S. Geological Survey Fact Sheet 2008-3082. Menlo Park: U.S. Geological Survey. Accessed July 26, 2012: http://pubs.usgs.gov/fs/2008/3082/.

Williams, C. F., Reed, M. J. & Mariner, R. H. (2008b). "A Review of Methods Applied by the U.S. Geological Survey in the Assessment of Identified Geothermal Resources." U.S. Geological Survey Open-File Report 2008-

1296. Accessed July 2, 2012: http://pubs.usgs.gov/of/2008/1296/pdf/of2008-1296.pdf.

World Bank. (30 June 2010). "Europe and Central Asia: Geothermal Energy Development Program (GeoFund) Project (English)." Accessed June 30, 2012: http://documents.worldbank.org/curated/en/2010/06/12820946/europe-central-asia- geothermal-energy-development-program-geofund-project.

World Bank. (n.d.). "Geothermal Energy Development Program, 2nd Tranche." Projects and Operations website. Accessed June 20, 2012: http://www.worldbank.org/projects/P094207/geothermal-energy-development-program- geofund-2nd-tranche?lang=en.

WSGR (Wilson Sonsini Goodrich & Rosati, PC). (2009). "FIPP, FIPP, Hooray…?: Analysis of and Commentary on DOE's Financial Institutional Partnership Program." Accessed November 6, 2012: http://www.wsgr.com/publications/pdfsearch/wsgralert_fipp.pdf.

End Notes

[1] This number is the combined mean estimate of identified and unidentified geothermal resources.

[2] In the United States, debt financing is rarely used for this phase of development (Salmon et al. 2011).

[3] The IFC (2013) defined successful wells as those with a potential of 3 MWe or greater, regardless of the potential return on investment of building a power plant at the site. Wells with a potential less than 3 MWe were also considered successful if connected to a currently operating power plant.

[4] See Figure 1 for a definition of Phases I, II, III, and IV. When a developer gains access to a geothermal resource but has not yet completed the steps to qualify as a phase I, this is termed as a "prospect." Note that the 2008 and 2009 GEA surveys did not report the project phases.

[5] Ormat Technologies put into service the first grid-connected geothermal plant powered by an enhanced geothermal system in April 2013 (EERE 2013). The plant has an estimated capacity of 1.7 MW.

[6] Traditional commercial lines of insurance, such as property and liability policies, do not necessarily function as credit enhancements. These policies are required as per the project's contracts (e.g., feed-in tariff and financing agreement), and development will not proceed without them in place.

[7] Additional policies implemented at the federal level in the 1970s to support geothermal (and other renewable energy technologies) included research, development, and deployment funding; the Public Utilities Regulatory Policies Act; and an investment tax credit (Doris et al. 2009).

[8] A project is operational once it is online and feeding power into the grid or directly to an end-user at the full level of intended capacity.

[9] For more information on these and additional exploration programs, such as the Geothermal Resource Exploration Development and Cascades Cost-Share, see: http://www1.eere.energy.gov/geothermal/pdfs/geothermal_history_1_exploration.pdf.

[10] The reason the 1703 program has not supported geothermal projects is likely because conventional geothermal projects are not considered "innovative clean technologies" as required by the program guidelines.

[11] This Act provided several other important programs to support the U.S. geothermal market. Title I included a provision for the "Geothermal Energy Coordination and Management Project to conduct resource inventory and assessment; conduct research and development; initiate a program to design and construct geothermal demonstrations plants; and provide for scientific and technical education programs through the National Science Foundation" (Nasr 1978). In addition to the loan guaranty program, "Title I provides for the establishment of … assistance in the payment of interest charges, and establishment of the Geothermal Resources Development Fund" (Nasr 1978).

[12] Although the reason for including the "full faith and credit of the United States" clause in the amendment is not known, it was likely done to indicate to lenders and investors the unconditional guarantee that would be provided by the federal government in the event of borrower default.

[13] The coverage ratio is the amount of funds allocated per loan to cover one potential failed loan; it represents one way to understand the perceived risk profile for a given loan guarantee program. The coverage ratio is akin to what is known as the "credit subsidy cost" under the present-day loan guarantee programs. The credit subsidy can be covered either by appropriations or a fee paid by the borrower (DOE LGP 2012).

[14] Information could not be found concerning when the last guarantee was made or exactly when the program ended.

[15] Section 1703 Title XVII of the 2005 *Energy Policy Act* provided for the Section 1703 Loan Program, which allows the federal government to ensure the repayment of loans for qualifying "innovative" clean technology projects, such as geothermal. However, only new technologies that have had "no more than three implementations that have been active for more than five years" are eligible. This requirement would seemingly make most geothermal projects ineligible for the 1703 program.

[16] It is not clear in any of the available program literature why the Geofund and ARGeo programs were discontinued or delayed.

[17] Targeted countries included Albania, Armenia, Azerbaijan, Belarus, Bosnia-Herzegovina, Bulgaria, Croatia, Georgia, Kazakhstan, Kyrgyzstan, Macedonia, Moldova, Montenegro, Romania, Russia, Serbia, Ukraine, Tajikistan, Turkey, Turkmenistan, and Uzbekistan.

[18] Projects in Ethiopia, Eritrea, Kenya, Rwanda, Tanzania, and Uganda are eligible.

[19] Projects in Ethiopia, Kenya, Rwanda, Tanzania, and Uganda are eligible.

[20] HS Orka is the only private firm that operates geothermal power plants in Iceland (Íslandbanki 2010).

[21] In some cases, NEDO and AIST transferred proven resources to the contracted industry partner in order for them to develop the resources for power generation; in such cases, the cost-share for exploration drilling applied (GeothermEx 2010). We think they may have tendered some of these proven resources as well.

[22] According to an interview conducted for this report, the GFF may no longer be completely "revolving" due to the high risk of geothermal exploration (Wahjosoedibjo 2012). That is, because losses are potentially high, fund designers thought the fund might be drawn down faster than it could be recapitalized.

[23] Most notably, Senator Jeffry Bingaman and Congressman Christopher Van Hollen introduced separate bills in 2009 that would have established a clean energy financing pool. Included in the American Clean Energy Leadership Act of 2009 (S.1462), Senator Bingaman proposed the Clean Energy Deployment Administration within the Department of Energy to provide financing for clean energy projects. Congressman Van Hollen introduced H.R. 1698 to establish a federally owned, independent "Green Bank" funded by U.S. Treasury bonds.

[24] In 2011 the State of Connecticut established the Clean Energy Finance and Investment Authority with the purpose of leveraging private investment in renewable energy and

energy efficiency projects. See www.ctcleanenergy.com for more information. California has two "green bank" institutions as well; see: http://www.treasurer.ca.gov/greenbank/index.asp. If these types of quasi-government "green banks" are established in states with geothermal resources such as California, they could conceivably provide low-cost financing for geothermal projects.

[25] Some municipal utilities, such as the Los Angeles Department of Water and Power, which serves nearly 4 million customers, are quite large. For more information, see: https://www.ladwp.com/ladwp/faces/ladwp/aboutus/a-water/a-w-factandfigures?_adf.ctrl-state=1653rwexct_21&_afrLoop=147738510606000.

In: Geothermal Exploration ...
Editor: Leon Edwards

ISBN: 978-1-63321-824-6

Chapter 2

GEOTHERMAL FIT DESIGN: INTERNATIONAL EXPERIENCE AND U.S. CONSIDERATIONS*

Wilson Rickerson, Jason Gifford, Robert Grace and Karlynn Cory

LIST OF ACRONYMS

BMU	Ministry for the Environment (Germany)
BMWI	Ministry of Economy and Finance (Germany)
CHP	combined heat and power
CREST	Cost of Renewable Energy Spreadsheet Tool
D/E	debt to equity
DSCR	debt service coverage ratio
EEG	Erneuerbare-Energien-Gesetz
EGS	enhanced geothermal systems
EPAct	Energy Policy Act
EU	European Union
FERC	Federal Energy Regulatory Commission
FIT	feed-in tariff
GW	gigawatt

* This is an edited, reformatted and augmented version of technical report, NREL/TP-6A20-53320, issued by the National Renewable Energy Laboratory, August 2012.

IPP	independent power provider
ITC	Investment Tax Credit
kWh	kilowatt-hour
LCOE	levelized cost of energy
MW	megawatt
NEA	National Energy Act
NREL	National Renewable Energy Laboratory
O&M	operation and maintenance
PPA	power purchase agreement
PTC	Production Tax Credit
PURPA	Public Utility Regulatory Policies Act
QF	qualifying facility
REC	renewable energy certificate
RFP	request for proposal
ROE	return on equity
RPS	renewable portfolio standard
RTO	regional transmission organization
SO4	Standard Offer No. 4
USGS	U.S. Geological Survey
WACC	weighted average cost of capital

EXECUTIVE SUMMARY

Developing power plants is a risky endeavor, whether the plants utilize conventional or renewable generation technologies. Feed-in tariff (FIT) policies can be designed to address some risks specific to geothermal power plant development, although they have not been used in the United States to date.

By April 2012, there were 11,224 megawatts (MW) of geothermal electric capacity installed in 24 countries (Holm 2010; GEA 2012). The United States leads the world with 3,187 MW installed. Despite these statistics, the cumulative amount of geothermal electricity generating capacity installed in the United States has expanded slowly in recent years, increasing approximately 2.3% annually on average since 2008 (GEA 2012). During 2011 and the first quarter of 2012, the addition of two new geothermal plants and the expansion of three existing plants added 91 MW of new capacity.[1] Geothermal market development faces a range of unique barriers related to the cost and uncertainty of resource exploration and confirmation drilling. In addition, geothermal projects face risks similar to other generation project development, including finding buyers for power, ensuring adequate transmission capacity, competing to supply electricity and/or renewable energy certificates (RECs), securing reliable revenue streams, navigating the legal issues related to project development, and reacting to changes in existing regulations or incentives.

Instead of addressing all these risks and all the policies in the same analysis, this analysis focuses on the design of FIT incentive policies for geothermal electric projects and how FITs can be used to reduce risks other than the physical risk of drilling unproductive exploratory wells. The guarantee of a stable revenue stream over the life of the generation plant lowers the risk that large investments in exploration and development will not yield an off-take agreement. The combination of a guaranteed purchase, a pre-determined payment price, and a standardized off- take agreement can relieve some of the cost, risk, and pressure associated with overall project development since the project does not need to compete for or negotiate a contract before final project costs are known. Addressing exploration risk[2] by incorporating some/all of the drilling costs in the FIT payment level can incentivize some amount of risk taking in the exploration phase. Policy risks should be considered too. The greatest risk associated with FITs—rapid and dramatic development in a timeframe too short to react and adjust—appears far less likely for geothermal projects due to: (1) longer development lead times and (2) geographic limitations on project locations. In short, policymakers should be able to see a problem coming with plenty of time to adjust. In sum, FIT policies can include risk-mitigating elements and can be tailored to support a few key financial risks of geothermal development.

Global Experience with FIT Incentives for Geothermal Electricity

FITs are currently the most common national renewable energy policy in the world and are in place in over 50 countries. Despite international interest in FITs, there has been a limited focus on geothermal FIT design specifically. This report represents the first in-depth attempt to explore how these policy elements could be aligned to support geothermal development. There are currently no geothermal FITs in North America, although the Public Utility Regulatory Policies Act (PURPA) of 1978, which contains elements similar to FIT policies, supported over 1,300 MW of geothermal development by the end of the 1990s (Guey-Lee 1999). Around the world, 16 countries have established specific FIT incentives for geothermal. Based on a survey of international experience to date, we find that:

- More than a dozen countries are experimenting with FIT policies that are specifically tailored for geothermal technologies. Countries have introduced FITs that target small scale geothermal plants, combined heat-and-power geothermal applications, specific types of geothermal resources (e.g., enhanced geothermal or low-temperature plants), specific geographies (e.g., islands), and domestic content.
- Geothermal FIT payment levels vary widely across countries from $0.08/kilowatt-hour (kWh) in Uganda to $0.48/kWh in Switzerland (both for 20-year contracts). This range reflects resource differences, variations in local electricity prices, as well as different policy objectives (e.g., whether to support certain resource types) in different countries and different approaches to rate setting.
- There is a lack of empirical evidence about the impact of the varied FIT designs on market development. Geothermal market growth under FITs has been slow to date for a variety of reasons. These include the fact that many of the FITs are comparatively new, and that some of the prices are too low to support the comparatively poor quality, or expensive local resources. A key take-away from the international survey is that the presence of FITs alone has not driven results comparable to the results achieved for wind and solar power under European FITs, or to those achieved for geothermal under PURPA in the United States.
- A variety of rate-setting approaches are currently used.
    - Examples in Germany and Spain demonstrate that the FIT rate-setting process can be used to reflect technology-specific risks and costs.
    - Different approaches to FIT rate setting have been taken in different countries.

FIT Policy Design to Support Geothermal Development in the United States

The report provides an in-depth, qualitative discussion on how FITs can be designed specifically to support geothermal power plant development in the

United States. For comparative purposes, this report adopts investor security as a perspective for discussing each option in order to analyze the relationship between policy design and risk reduction.[3]

The report identified key issues relating to geothermal FITs in the United States, listed in Table ES-1 below. Each policy design issue is evaluated based on its potential to lower investor risks.

Table ES-1. Summary of Geothermal FIT Design Issues and Options

Policy Design Issue	Design Option Potential for Lowering Investor Risk
Integrating FITs with Geothermal Targets	Setting mandatory and specific targets for geothermal can improve investor confidence in a government's commitment to geothermal procurement.
Technology Eligibility and Differentiation	The choice of eligibility or differentiation does not have direct impact on project risk. What matters from an investor risk perspective is whether the FIT rate available to geothermal generators of various sizes and types is sufficient to cover operating expenses and meet investors' required returns.
Incorporating Exploration Expense into FIT Rates	Accounting for exploration costs and risk when setting the FIT rate (i.e., by building exploration cost and probability of success assumptions into the rate setting model) can be an avenue for reducing investor risk and encouraging the flow of capital into geothermal development. Assurance of a financeable contract that enables the recovery of an estimate of incurred exploration expenses provides a signal to the market to invest, notwithstanding the fact that actual exploration experience is likely to differ from the estimate and that some exploration efforts will fail to produce a viable thermal resource.
Standard Pricing and Off-Take Contracts	FITs provide generators with standardized off-take contracts that can reduce power project risk and transaction costs. This has the potential to reduce investor return requirements and increase the pool of available capital.
Payment Structure	Fixed FIT prices are lower risk (from the investors' perspective) than payment streams that vary over time and are uncertain (Corfee et al. 2010).
Contract Duration	To the extent that post-contract market revenue projections are utilized in setting a FIT payment level, longer contract

Table ES-1. (Continued)

Policy Design Issue	Design Option Potential for Lowering Investor Risk
	durations reduce investor risk by reducing the proportion of lifetime revenues exposed to market price risk. If, however, FIT payments are designed to attain target investor returns prior to contract termination, investors should be indifferent to contract duration, all else being equal.
Policy and Contract Timing	The presence of a stable FIT policy that could be relied on for a period of several years by an early-stage, exploration-phase geothermal investor could decrease overall risk, thus reducing/eliminating developer risk premiums for request for proposal (RFP) bids.
Payment Indexing Over Time	Indexing payments to inflation, drilling costs, and/or other relevant indicators reduces the risk that project revenues will be inadequate.
Adjusting the Policy	Adjustments that are made infrequently and on a schedule that is known in advance and using a process that is as transparent as possible increases investor confidence.

Analyzing the Impact of Policy Design Choices on the Levelized Cost of Energy from Geothermal Electricity Generators

This report provides a quantitative analysis using the National Renewable Energy Laboratory's (NREL) Cost of Renewable Energy Spreadsheet Tool (CREST) model to illustrate policy design impacts on the levelized cost of energy (LCOE) for geothermal electricity. The purpose of this analysis is not to determine the precise cost of a specific geothermal system. Rather, the analysis identifies considerations that can inform policymakers about how different FIT policy design choices can address some of the geothermal-specific risks that contribute to the LCOE, and thus how FIT design can be used to lower the LCOE of geothermal energy.

The quantitative analysis focuses on the interaction of FITs with several key variables:

- Exploration success rate
- Duration of the confirmation stage

- Investor return requirements (equity and debt, if applicable) during each phase of development
- FIT contract duration
- The project's total installed cost (including operations and maintenance costs).

The results show a connection between the investor return requirement, the amount of time it takes to complete a geothermal project (the amount of time this capital is deployed), and the LCOE. The impact that long-term, price-certain contracts with creditworthy counterparties have on project financing is also explored. Such contracts can help make project-level equity and debt more accessible and on more favorable terms. FIT designs that reduce revenue risk and contract price risk can reduce the associated investor risk and return requirements at each development phase. The availability of a long-term, price-certain FIT can shorten the geothermal development cycle by removing the need to compete for a power purchase agreement (PPA), which can also translate into a lower LCOE. Longer FIT contract durations can also achieve a lower LCOE. With regard to the impact of policy design on LCOE, this analysis shows that policies that reduce the duration and return requirements of early-stage development activities will have the greatest impact. Increasing geothermal contract lengths and introducing mechanisms that reduce the cost of permanent equity and debt also have a significant impact on LCOE. The approach to setting payment levels under FIT policies (e.g., whether and how exploration and confirmation success and return requirements are taken into account when calculating payments) is also shown to have a direct and meaningful impact on geothermal LCOE.

1. INTRODUCTION

Geothermal electricity[4] represents a significant source of potential renewable electricity in the United States and it has several comparative advantages over other types of generation. Geothermal energy is a renewable resource that provides baseload, rather than intermittent, power at high availability and capacity factors (i.e., at or above 90%) (Green and Nix 2006). Geothermal power is a cost-effective alternative to non-renewable energy sources, primarily in western states where the strongest resource is concentrated, and has been used to effectively as a strategy to diversify state electricity portfolios since 1960 (Mines et al. 2010). However, developers of hydrothermal geothermal power plants face a unique and added challenge compared to other renewable energy technologies. In addition to securing sites, permits, and contracts (all required to attract financing), geothermal developers also must find and confirm a commercially-exploitable thermal resource. The expense of drilling, and the potential to expend substantial capital without yielding a developable well, leaves geothermal developers with substantial exploration risk. Because of such project uncertainties, many geothermal resources remain undeveloped. The unique characteristics of geothermal electric development have important implications for effective policy design of geothermal projects.

This report explores how FITs have been, and can be, designed to account for the unique characteristics of geothermal project development, particularly reducing investor risk in the power plant. NREL is currently working on a separate report that explores policies to encourage geothermal exploration. This report explores international experience with geothermal FITs to date and discusses design considerations for geothermal FITs in the United States. The report uses NREL's CREST[5] to quantitatively examine the impacts and tradeoffs of different geothermal FIT designs. The analyses explore, for example, how different FIT designs can take development risk into account and the impact that reducing the cost of capital by decreasing investor risk could have on project economics. The report is organized as follows:

- Section 2 provides a high level summary of U.S. geothermal energy, including geothermal resources, market trends[6], and policies. The section also reviews the barriers and risks that constrain geothermal energy development and the potential for different FIT design options to overcome them.
- Section 3 defines FIT policy characteristics, reviews international experience with geothermal FITs in 16 countries (none in North America), and compares current FIT payment levels in these countries. Examples of the geothermal payment level setting process are detailed for Germany and Spain.
- Section 4 provides an in-depth, qualitative discussion on how FITs can be designed to address U.S. considerations for geothermal energy development.[7] For comparative purposes, this section adopts investor security as a benchmark for discussing each option, but there is great flexibility in how geothermal FITs can be structured, designed, and evaluated.

- Section 5 presents a quantitative analysis of the design tradeoffs identified in Section 4. Specifically, this section utilizes the CREST model to illustrate the impact of various policy design choices on LCOE. These include the impact of strategies to reduce exploration, permitting, and financing risks.
- Section 6 contains conclusions and discussions of key findings.

2. Overview of Geothermal Development Considerations and Risks

2.1. Overview of the U.S. Geothermal Market

The United States has a significant presence in the global geothermal energy market. As of April 2012, there are 11,224 MW of geothermal electric capacity installed worldwide (GEA 2012). The United States has the highest amount of geothermal capacity, with 3,187 MW installed (Figure 1). Of an estimated $2 billion in new financial investments made in geothermal energy in 2010, $0.7 billion, or 35%, was invested in the United States (McCrone et al. 2011).[8] Despite these statistics, the cumulative amount of U.S geothermal electricity installed has expanded slowly in recent years, increasing an average of 2.3% each year since 2008. From 2011 through the first quarter of 2012, two new projects and three expansions at existing facilities added 91 MW of geothermal capacity in the US (GEA 2012).

The growth of the geothermal market stands in contrast to the U.S. wind and solar markets, which have grown significantly during that same time period (EERE 2011). Figure 2 shows annual U.S. electricity production for wind and geothermal over the last decade. In 2004, for example, the amount of electricity generated by geothermal power plants was slightly higher than the amount of electricity generated by wind energy. By 2011, however, the total amount of wind generation was over seven times higher than the total amount of geothermal generation.

Geothermal growth has been slow compared to other renewable technologies, but there is significant potential to expand the market. The section below briefly reviews the U.S. geothermal resources available for electricity generation.

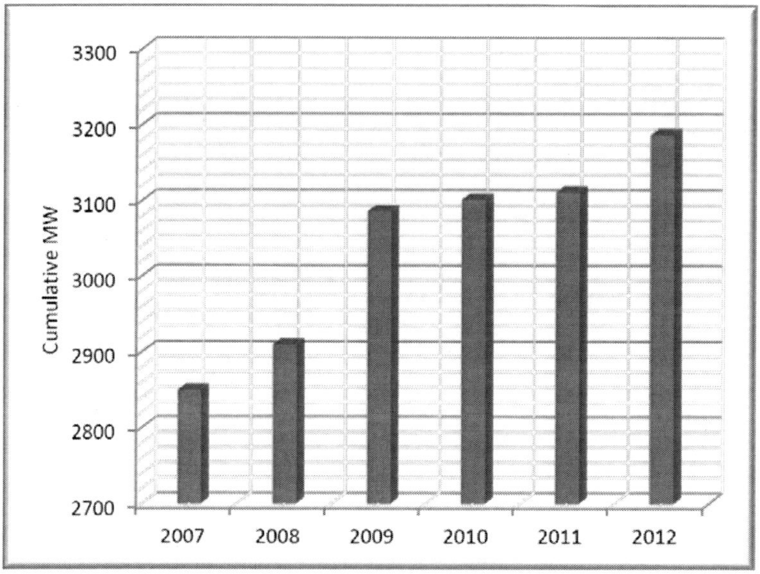

Source: GEA (2012).

Figure 1. Cumulative installed U.S. geothermal electricity capacity (2008-First Quarter 2012).

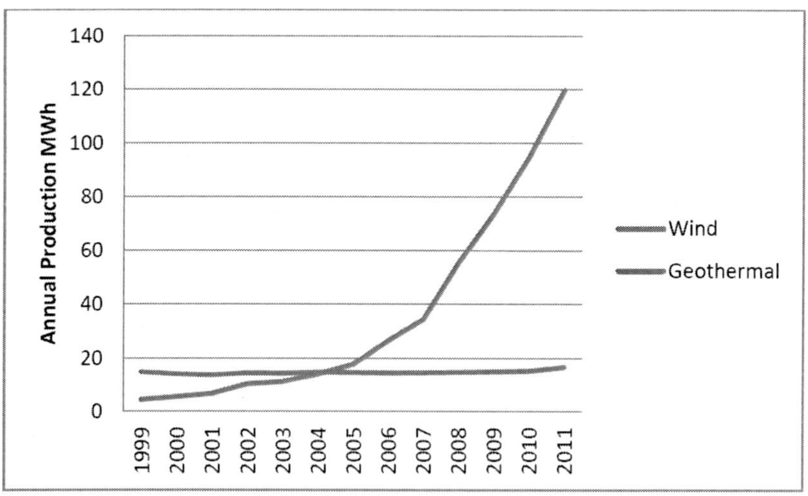

Source: U.S. EIA (2012).

Figure 2. U.S. electricity generated by wind and geothermal plants (1999-2011).

2.2. U.S. Geothermal Resources

The exact amount of available geothermal energy is unknown, but recent estimates have demonstrated that U.S. geothermal potential remains vast and largely untapped. The different types of potential resources include[9]:

- Hydrothermal resources. Hydrothermal systems are naturally occurring underground geothermal reservoirs. Hydrothermal reservoirs can either be dominated by hot water or steam.[10] Hydrothermal reservoirs are continuously recharged as heat flows into the reservoir from greater depths. If identified hydrothermal resources were developed to their full extent, the U.S. Geological Survey (USGS) estimates that that an additional 6.5 gigawatts (GW) of geothermal electricity capacity could be brought online across 11 western states, Alaska, and Hawaii (Williams et al. 2008). The USGS also estimates that there is an additional 30.3 GW of unidentified hydrothermal resource.
- Deep or enhanced geothermal systems (EGS).[11] EGS resources are engineered reservoirs that can be created by fracturing rock deep in the earth. Water is then injected into the fractures, heated through contact with the rock, and re-circulated to the surface. The Massachusetts Institute of Technology (MIT) estimated that the extractable portion of available EGS heat energy would exceed 2,000 times the annual consumption of U.S. primary energy in 2005 and that 100 GW could be developed by 2050 if sufficient investment were committed (MIT 2006). The USGS, meanwhile, estimates the available EGS resource to be 500 GW (Williams et al. 2008).[12]
- Co-produced resources. Hot water is often pumped up from underground as part of the oil and gas exploration process. This water could be captured and harnessed to produce geothermal energy in small-scale power plants connected to the distribution network (McKenna and Blackwell 2005; Hurlbut 2012). The total amount of co-produced geothermal resources is uncertain and has not been recently or comprehensively studied on a nationwide basis (Augustine et al. 2010).

Today, the vast majority of geothermal electricity generation comes from hydrothermal resources, a minimal amount from co-produced resources, and only experimental plants use EGS. Given these potentially available resources, there is significant potential for the United States to scale-up its production of geothermal electricity. Market development, however, has been slow because of the presence of project risks and obstacles, as described below.

2.3. Obstacles to Geothermal Power Plant Development

Geothermal generation project developers face many of the same types of risks that any power plant project will face, such as:

- Development Risk (Timing). The project will not achieve milestones (e.g., permitting, siting, and interconnection), will be delayed, and/or not completed.
- Development Risk (Contracting). The renewable energy project investor incurs significant cost on site control, engineering, permitting, and other development activities, as well as legal and other costs associated with competing for, or negotiating, a PPA, prior to knowing whether it will successfully secure a PPA, RECs and other commodities with a creditworthy buyer (i.e., off-taker). Investments in such development-stage are fully at risk, and therefore have commensurate return requirements.
- Contract price risk. The PPA price, which may be set (or proposed in a competitive solicitation) based on preliminary estimates and committed to before project development activities are complete and project costs are fully known, may not be sufficient to cover actual costs.[13]
- Construction risks. The project will be delayed because of slow construction or negatively impacted because of construction costs overruns.
- Technology risk. The technology does not perform as expected.
- Resource risk. The renewable energy resource is not as strong as projected.
- Market revenue risk. The revenue received by the project may be volatile over time, which may negatively impact project economic performance.
- Political (regulatory or legislative) risk. The policy framework under which the project was developed may be altered, which may negatively impact project revenues or operations.

These risks can constrain project development and increase the cost of capital to finance projects if not adequately addressed. For geothermal projects, the project risks outlined above can be magnified or compounded by the characteristics of the geothermal exploration and development process:

- Development timetable. The length of time to develop a geothermal project is longer than other renewable energy generators and is typically 4-7 years (NREL 2011; Richter 2009). The development process is complex and involves several distinct stages: exploration, confirmation drilling, production drilling, and power plant construction. The duration and complexity of these activities creates significant opportunity costs compared to the development of other technologies with shorter development cycles and greater opportunities for success.
- Complexity of siting and permitting. The geothermal siting and permitting process can be challenging because of a lack of clarity in resource ownership rights and because geothermal development may require complex interactions with overlapping federal and state regulatory bodies (Doris et al. 2009; Fish and Heaps 2009).
- Transmission capacity. The best geothermal plants are often located far from large load centers. Geothermal power plants are faced with the choice of either using existing transmission capacity, which is often constrained, or attempting to build new transmission capacity, which is typically prohibitively expensive to build to serve a single power plant. A key challenge with new geothermal development, therefore, is identifying the transmission infrastructure that will bring the power to market (Hurlbut 2012).
- Development and exploration costs. The upfront costs of developing geothermal projects are substantially higher compared to other energy generation technologies, and can account for 40%-50% of total geothermal project costs in the United States. (Cross and Freeman 2009; Deloitte 2008).
- Exploration risk. While the development process is expensive, it is not guaranteed to be successful; geothermal projects have an exploration success rate of 10%-40% (Augustine et al. 2010; Hance 2005). In other words, geothermal prospecting may incur significant costs from surface exploration and exploratory drilling and yield only "dry" wells. (Importantly, the actual probability of dry wells depends on the hydrothermal application.[14]) This contributes to the upfront cost;

drillers need to make higher returns on each successful well to make up for the dry wells. As a result of these risks, geothermal exploration may be supported by government programs or by industry participants with substantial balance sheets and a long-term outlook on geothermal electricity markets. In either case, this capital can be very difficult to obtain (Richter 2009; Salmon et al. 2011).

2.4. Addressing Geothermal Barriers and Risks through Policy

The combination of the duration, complexity, cost, and other risks inherent of geothermal development means that geothermal power plants are challenging to build and finance. Geothermal projects typically require policy support in order to decrease project risks and overcome barriers.[15] Policies can be used to decrease the risk and complexity of geothermal project development and financing.

There are numerous policy tools that can be used to reduce exploration and development risk and improve access to capital. These could include improved research and development, government risk sharing (e.g., publicly-funded or managed exploration, co-investment, drilling insurance, etc.), the development and coordination of inventories of geothermal data, and inter-agency coordination (Deloitte 2008; Doris et al. 2009; GEOFAR 2009a). This report does not focus on policies geared to directly address exploration risk since these were explored in other research efforts, including a forthcoming NREL report.

This report instead focuses on analyzing policy options that establish revenue certainty and ensure known access to a revenue stream if exploration proves successful, removing not only the transaction, bid security, and legal costs of pursing a PPA, but mitigating a variety of the aforementioned risks.[16] Decreasing risks can increase investor confidence and reduce the time for financing. Recent studies have estimated that targeted renewable energy policy can reduce risk, lower the cost of capital required to finance wind and solar projects, and result in a 10%-50% decrease in the levelized cost of energy of renewable generation (Varadarajan et al. 2011; de Jager and Rathmann 2008). This report focuses on policies that can address the following risks:

- Market revenue risk. Market revenue risk occurs when the price paid to the generator varies over time or when the future availability of a revenue stream (or other cash- equivalent benefit stream) is unclear.

Reliance on spot electricity markets for revenues is one form of market revenue risk. The temporary nature of federal incentives, including the Production Tax Credit (PTC) for example (i.e. reauthorized for 1-4 years at a time), can also create a net cash flow risk for geothermal investors and developers with long development timelines (Salmon et al. 2011).[17] Market revenue risk can be diminished by policies that provide geothermal developers with stable, long-term payments that are known in advance. Alleviating this risk can significantly improve investor security and reduce the cost of capital (Deutsche Bank Climate Change Advisors 2009).

- Development risk. Development risk reflects the fact that a project may or may not find a buyer for its power, RECs, and/or other commodities before expending significant development capital on siting, engineering, and permitting. This could result, for example, from the absence of a policy stimulating demand for geothermal power, shortage of viable counterparties, loss of a competitive procurement, or failure to conclude a successful power purchase agreement negotiation with a prospective buyer. This risk can be reduced through policy by assuring that a project can rely on access to a PPA if it can find a technically and economically viable resource and secure necessary permits.
- Contract price risk. Even when a mechanism is in place to provide a geothermal developer with an opportunity to compete for a PPA (such as under an renewable portfolio standard (RPS) in a western U.S. vertically-integrated electricity market), it is common that developers may need to offer price proposals before development contingencies have been fully resolved and therefore before project costs are fully understood. As a result, developers typically either build in a risk margin to account for these risks, or incur additional expenses to attempt to hedge uncertain costs. Alleviating the need to build in risk premiums or hedging costs can reduce the viable PPA price (Wiser et al., 2006).[18] From the policymaker's perspective, this risk may be daunting –how does one set the "right" payment level? Some countries, like Switzerland, require that developers participating in the FIT provide cost data so that they have a database of cost information that grows over time and can be used to inform future payment levels (Couture et al. 2010).
- Legal risk. Legal risk refers to the risk of legal delays, costs, and failures during the process of competing for and negotiating a power

purchase agreement. These risks can be diminished by policies that provide incentives that are available on a standard offer basis under standardized contracts.

- Political risk. Exposure to policy changes impacts a geothermal developer's prospects and cannot be hedged by project investors. The longer development horizons of geothermal power leave investors more exposed to such risks than investors in wind or solar power. These risks can be lessened by policies that provide incentives that are available on a standard offer basis under standardized contracts, particularly if such incentives are established for a sufficient duration to allow for geothermal development cycles. Buyers often seek protection from such risks under PPAs, which can shift substantial risk to project owners. Standard PPAs without such political 'outs' can therefore reduce investor risk.

Several recent studies have surveyed non-exploration geothermal policies in the United States and evaluated their adequacy for mitigating risks, addressing barriers, and driving market development of new (or incremental) geothermal electricity facilities. NREL (Doris et al. 2009), for example, reviewed state policy and concluded that:

- With few exceptions, current renewable energy policies are not sufficiently tailored to provide new geothermal power plants with the specific types of support they require. RPSs for new generation support inter-technology competition, which, in practice, has primarily favored new wind energy for reasons that may span differences in resource availability, development timeline, and exploration/development risk. By 2010, 91.2% of the new renewable energy capacity motivated by RPS policies was wind, whereas only 1.4% was geothermal (Barbose 2011). These specific values are influenced by the fact that state RPS programs are distributed nationally but geothermal development has been thus far limited to the western United States.
- The policies that specifically support geothermal are often limited in their ability to meaningfully impact the geothermal market due to funding or eligibility (e.g., project size) restrictions.

In markets such as California's where geothermal power plays a material role in RPS compliance, the policies are structured to include existing resources in utility RPS procurements.[19] In many cases, these geothermal resources were originally developed under the PURPA. The feasibility of expansions at existing geothermal facilities has also been investigated as a result of the long-term utility procurements associated with RPS programs. In this case, the potential availability of a long-term contract provides the opportunity to develop a financing plan for expansion, but the short lead-time afforded by the RFP process presents a continuing challenge to long-lead time resources like geothermal. This issue is compounded for geothermal developers seeking financing for greenfield development.

The NREL study identifies FITs as one possible policy approach that could expand or improve the existing policy landscape.[20] FITs have been identified in studies by Deutsche Bank and others as one of the most effective policies for mitigating key renewable energy project development risks, attracting investment capital, and scaling-up renewable energy markets (Deutsche Bank Climate Change Advisors 2009). While experience is early, FITs can be designed to target geothermal generators and reduce geothermal project risks; specifically, geothermal FITs can be structured to:

- Pay generators with a fixed, long-term price for electricity, RECs, and/or other commodities, which can decrease market revenue risks. Projects do not need to compete for or negotiate a contract before final project costs are known, reducing contract price risk. Projects under such circumstances may also not be subject to contractual milestones that could lead to contractual penalties, forfeiture of security payments, and termination of the off-take agreement if there are project delays.
- Require utilities (or other entities) to purchase geothermal output on a standard offer basis, which can ensure contract access. This removes the risk that investments in exploration and development will not yield an off-take agreement. This certainty may also attract a broader pool of capital providers to finance both the exploration and operations periods, thereby potentially lowering capital costs.
- Provide generators with standardized contracts that can reduce transaction costs.

FITs might help the U.S. geothermal market achieve accelerated expansion goals, as a complement to policies designed specifically to address exploration risks. FITs that address revenue, market, development, legal, and political risks can have a positive impact on the decision of geothermal developers to move forward with geothermal exploration by removing some of the uncertainties of the development process in advance and accelerating development times.

There are, however, limitations and challenges with FIT policies. First, FIT payments are typically set administratively rather than through competitive processes. Similar to tax credits, rebates, and performance-based incentives, policymakers need to carefully set the "right" price that most fully meets their FIT policy objectives. Setting a price too low may yield less development than desired whereas setting the rate too high may lead to overpayment and more development than desired. Second, without a competitive mechanism to help determine or benchmark incentive levels, the setting of FIT prices typically requires intensive data gathering processes. Third, periodic reviews are necessary to help ensure that the payment levels reflect market conditions and policy objectives accurately. Finally, some FITs have been designed without caps on the amount of generation to be developed (e.g., Germany). This approach can create uncertainty with regard to total policy impact and policy cost, and in some cases (involving short lead-time generation with ample resource potential), has led to substantial rate impacts (e.g., photovoltaics in Spain). In order to address these concerns, there are mechanisms that can be used to limit market growth and contain costs, such as caps and payment level adjustment mechanisms (Kreycik et al. 2011).

The remainder of this report focuses on geothermal FIT design considerations and the risk implications of different design options in greater detail. This report does not revisit or summarize broader discussions of how FITs compare and contrast with other policy types. The next section explores FITs by first defining what they mean and then examining their application in support of geothermal energy in other countries.

3. Global Experience with FIT Incentives for Geothermal Electricity

3.1. Defining FITs

Although there are no geothermal FITs in North America, there are currently 16 countries that have established specific FIT incentives for geothermal.

FITs, also referred to as *standard offers* or *CLEAN contracts,*[21] are currently the most common national renewable energy policy in the world and are currently in place in over 50 countries (REN21 2011). FITs have driven the rapid scale-up of renewable energy technologies around the world and have supported 87% of the global photovoltaic capacity and 64% of global wind energy capacity as of 2010 (Tringas 2011).

Several recent publications have catalogued and described the primary FIT design considerations (Couture et al. 2010; Grace et al. 2008; Klein et al. 2008; Mendonça et al. 2009). What is clear from these studies is that FITs are complex policies that can be implemented in a wide variety of forms. In fact, no two FITs have been implemented in the same way. The complexity of FITs is attributable to the fact that they are not a single policy focused on the payment level, but rather a package of regulations and incentives, which can contain all of the following elements:

- Interconnection rules. The term "feed-in" tariff derives from the 1991 law in Germany that first guaranteed independent power producers the right to connect to and feed their electricity into the grid. FIT policies may include requirements that utilities interconnect renewable generation ("guaranteed interconnection"), that renewable energy advance ahead of conventional generation in the interconnection queue ("priority interconnection"), and/or that the cost of interconnection or grid upgrades be passed through to ratepayers (Rickerson et al. 2011; Tweedie and Doris 2011).
- Purchase and dispatch requirements. FIT policies may also require that utilities not only purchase all of a generator's output[22], but then also prioritize its delivery ahead of non- renewable generation. If renewable generation must be shut down or curtailed by the grid operator for technical reasons (e.g., in order to ensure grid security), some FIT policies further guarantee that generators will receive payment for the electricity they were unable to sell[23] (Rogers et al. 2010).
- Standardized contracts. FIT policies may include provisions that require utilities to offer standard and simplified contracts to generators, rather than requiring generators to negotiate contracts on a case-by-case basis (Couture et al. 2010).
- Pre-determined long-term payments. FITs typically include a standard offer for a known payment to generators, available on a first come, first-served basis. FIT payment levels are usually set administratively,

rather than relying on competition or market mechanisms, and are typically offered for durations of 15 to 20 years (Couture et al. 2010).

- Other policy objectives. Additional FIT policy features could include clear transmission cost allocation, cost of system upgrades, cost caps that limit total policy cost, forecasting obligations for variable generation, bonus payments to encourage generation where the electric system most needs it (e.g., urban areas, radial ends of distribution systems), bonus payments for innovative technologies, and repowering of old sites (Couture et al. 2010).

Despite international interest in FITs, there has not yet been much research that focuses on geothermal FIT design specifically. This report represents the first in-depth attempt to explore how these policy elements could be aligned to support geothermal development. As a first step, geothermal FITs internationally were surveyed in order to identify lessons learned. While it is too early to mathematically quantify the impact of geothermal FITs on accelerating project development, some early observations provide some useful insights.

3.2. FITs in North America

FITs have only recently appeared in North America. Ontario established its current FIT policy in 2009 as a replacement for its earlier Renewable Energy Standard Offer Program (RESOP) (Ontario Power Authority 2011). Nova Scotia followed in Ontario's footsteps by establishing its Community Feed-in Tariff (COMFIT) in 2011 (NSUARB 2011). A number of U.S. states have introduced FIT legislation during the past several years, but only California, Hawaii, and Vermont have enacted FITs (Couture and Cory 2009; Rickerson et al. 2008). Additionally, several cities[24] have enacted FIT policies.

The experience with FITs in the United States and Canada, however, has been of limited relevance to geothermal either because geothermal is not eligible to participate, is not a viable local resource, or because the programs impose size limitations (e.g., for distributed, on-site generation) that would effectively exclude most geothermal electric generation.[25]

The closest analogue to large-scale FITs for sizable project in the United States is PURPA of 1978. Under PURPA, utilities were required to purchase power offered by "small power production facilities[26]" at the utilities' full avoided cost. The calculation of avoided cost was left to state discretion.

Under PURPA, California created the Standard Offer No. 4 (SO4) contract, which defined avoided cost as a 10-year schedule of escalating payments (Pierce and Livesay 1994). The GTP Blue Ribbon panel noted that SO4 "drove geothermal exploration in the 1980s…Private industry was willing to take more drilling risk because of the higher electricity price (EERE 2011)." In total, PURPA supported 1,346 MW of geothermal capacity, including 1,295 MW in California, 237 MW in Nevada, and 35 MW in Hawaii (Guey-Lee 1999). A more detailed discussion of PURPA and geothermal development can be found in Appendix C.

PURPA demonstrated that the combination of a guaranteed purchase and a fixed price could attract investment in geothermal development. PURPA was controversial, however, because the prices that were paid under long-term contract were judged to be too high in retrospect for many projects (Hirsh 1999).[27] The effectiveness of PURPA subsided as avoided costs levels decreased with the dramatic fall in natural gas and oil prices. The Energy Policy Act of 2005 (EPAct) further restricted PURPA by exempting utilities that allow non-utility generators access to competitive markets from having to sign new PURPA contracts upon a demonstration to the Federal Energy Regulatory Commission (FERC) (Elefant 2011). Since PURPA provided generators with guaranteed market access and, in the case of SO4, a fixed price for power, many analysts refer to PURPA as the world's first FIT (Lipp 2007; Mendonça et al. 2009; REN21 2011). It is important to note, however, that PURPA was not tailored specifically to support geothermal energy and questions remain as to whether a more targeted policy design could better decrease geothermal development risk while providing generators with lower payments.

3.3. International FITs

International experience with geothermal-specific FITs is limited, but lessons learned to date can inform policy design for geothermal electricity. This analysis of these FIT policies focuses on several key issues:

- A high-level comparison of FIT payment levels for geothermal across different countries
- The impact of geothermal FITs on the development of new capacity worldwide

- Approaches to account for development and exploration costs when setting the FIT payment level

In order to gather data on geothermal policies, reviews of original FIT legislation (or translated legislation) were conducted if readily available in English, Spanish, or German. When laws were not either readily available and/or un-translated, secondary sources that summarized policy content (EGEC 2011; Rickerson et al. 2010) and websites such as RES LEGAL[28] were consulted. In total, 16 countries were identified with FITs specifically for geothermal.[29]

3.3.1. FIT Payment Levels and Impacts by Country

The two tables below contain a high-level summary of the geothermal electric FITs surveyed. Table 1 summarizes geothermal FITs in European Union (EU) member states, whereas Table 2 summarizes geothermal FITs in non-EU countries. The following data is shown for each country:

- FIT payment structure refers to whether the payment is made as a fixed amount for electricity or whether it is paid as a premium on top of the wholesale price for electricity.
- Requirements for FIT payment. The types of geothermal plants that are eligible for FIT payments vary by country. Some of the countries specify constraints such as a minimum efficiency or a maximum project size, whereas other countries set forth multiple payment levels differentiated based on project size, resource temperature, or geographic location. Some countries allow bonus payments for plants with certain desirable characteristics (i.e., specific technology, ability to achieve commercial operation within a specified period of time). If a specific plant meets all of these characteristics, then it can claim all of the bonus payments, incremental to the base FIT payment.
- FIT payment duration refers to the number of years that a geothermal plant is paid the FIT incentive. This ranges from 10 years (e.g., Turkey) to 20 years (e.g., Germany and France).
- Geothermal FIT payment refers to the amount of money received for every kilowatt-hour generated under the FIT. This amount is listed in both Euro (€) and in U.S. dollars (US$). Currency conversion throughout this report is based on the Euro-USD exchange rate of 1.4 on November 1, 2011.

- Geothermal Cumulative Installed Capacity refers to the total cumulative installed geothermal electric capacity in the country.
- Cumulative Geothermal Capacity Supported by FIT indicates the total cumulative amount of installed geothermal electric capacity that receives payments under the FIT policy. As can be seen in the table, the amount of geothermal capacity supported by FITs to date has been limited and this is discussed in greater detail in Section 3.3.2.

3.3.2. Lessons Learned from International FITs

A high-level review of the policy design and implementation details of global geothermal FITs provides several insights:

- A diversity of geothermal FIT designs. It is noteworthy that of the approximately 50 countries that have enacted FITs internationally, approximately one-third have crafted specific incentive levels targeting geothermal development. Of these, each of the different countries has designed their FITs to reflect different policy objectives and geothermal resource characteristics. These include a focus on small scale projects (Italy), minimum plant efficiencies (Austria), specific resources (e.g., EGS Germany and low temperature in Greece), different geographies (e.g., islands vs. mainland in Ecuador and France), and domestic content (Turkey). These design considerations can serve as useful benchmarks as policymakers in other countries consider their own customized designs.
- A broad range of remuneration levels. The diversity of FIT designs is also reflected in the manner in which the FIT payment levels are set. Figure 3 compares the highest rate paid for geothermal in each country by calculating the net present value[34] of the revenue that would be paid to a one megawatt geothermal plant for the duration of the FIT contract as a means of comparing contracts of different lengths. This illustrates, for example, that the overall value of geothermal in the Czech Republic is higher than that in Croatia. Although both countries pay generators €0.18/kWh, generators receive this payment for 20 years in the Czech Republic instead only 12 years in Croatia.[35] As can be seen in the figure below, the value provided to generators varies widely. This reflects the fact that some countries have established payments designed to reflect the generation costs of comparatively expensive technologies, such as EGS in Germany or smaller-scale, island- based systems in France.[36] Other

countries, such as Serbia, have set payment levels more conservatively in order to target only lower-cost generators (Kolarevic 2009). The examples of Germany and Spain are discussed in greater detail in Sections 3.3.3.1 and 3.3.3.2.

- A lack of empirical data on FIT impacts. Although the range of different designs is a useful benchmark, there is currently a lack of data with which to compare the impact of different design decisions. Despite the success of PURPA in the United States (and SO4 in particular), response to international FITs to date has been limited. Only the FITs in Austria, Germany, and France have supported new geothermal capacity and, in these cases, the total amount of new generation has been under 20 MW. [37]

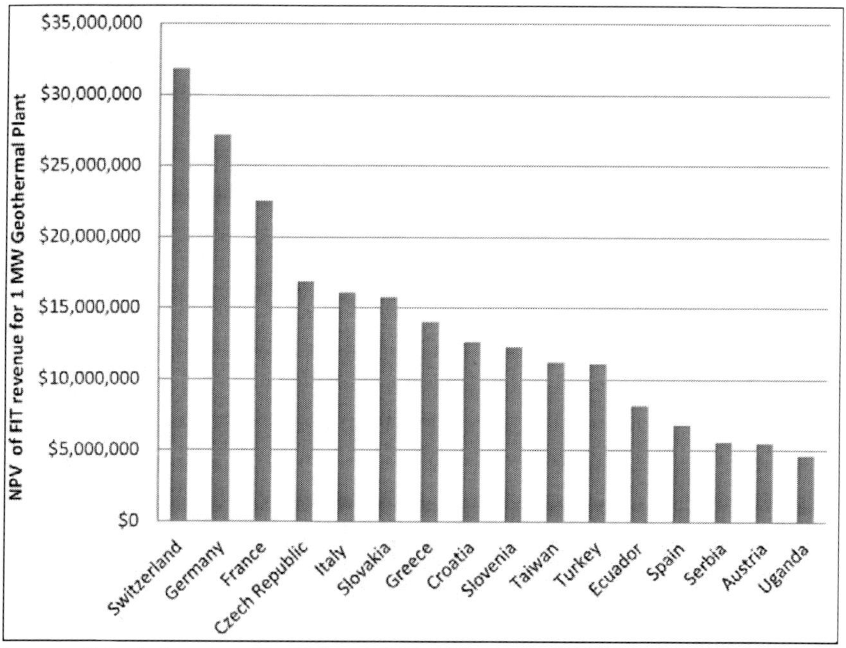

Source: Calculations based on Figure 1 and Figure 2 (72% net capacity factor, 7% discount rate).

Figure 3. Comparison of FIT revenue NPV for 1-MW geothermal plant.

Table 1. Geothermal Electric FITs in the European Union (EU)

Country	FIT Payment Structure*	Requirements for FIT Payment	FIT Payment Duration (Years)	Geothermal FIT Payment (€/kWh)	Geothermal FIT Payment (US$/kWh)	Geothermal Cumulative Installed Capacity (MW) (2010)	Cumulative Geothermal Capacity Supported by FIT (MW) (as of 2010)
Austria	Fixed	Minimum 60% power plant efficiency	13	0.075	0.11	1.4	1.2[30]
Czech Republic	Fixed	None	20	0.18	0.25	0	0
	Premium	None		0.15	0.21		
France	Fixed	Continental France (< 12 MW)	15	0.2	0.28	16.5	16.5
		Bonus payment for system efficiency (< 12 MW)		0.08	0.11		
		Overseas Territories (OT) (< 12 MW)		0.13	0.18		
		OT bonus for system efficiency(< 12 MW)		0.03	0.04		
Germany	Fixed	Base rate	20	0.25	0.35	7.5	7.5
		Bonus payment for EGS		0.04	0.056		
Greece	Fixed	Low temperature geothermal	20	0.15	0.21	0	0
		High temperature geothermal		0.099	0.14		
		Bonus payment if generator does not receive other subsidies		0.2	0.28		

Table 1. (Continued)

Country	FIT Payment Structure*	Requirements for FIT Payment	FIT Payment Duration (Years)	Geothermal FIT Payment (€/kWh)	Geothermal FIT Payment (US$/kWh)	Geothermal Cumulative Installed Capacity (MW) (2010)	Cumulative Geothermal Capacity Supported by FIT (MW) (as of 2010)
Italy [31]	Fixed	< 1 MW	15	0.2	0.28	843	0
Spain	Fixed	< 50 MW	20	0.073	0.10	0	0
Slovakia	Fixed	< 10 MW	15	0.196	0.27	0	0
Slovenia	Fixed	< 5 MW	15	0.153	0.21	0	0
	Premium	< 10 MW	15	0.093	0.13	0	0

* FIT payment structure refers to whether the payments are fixed over time or whether they are premium paid on top of wholesale electricity prices. Sources: (Bertani 2010; BMU 2011b; EGEC 2011; Holm et al. 2010; IGA 2011)[32]

Table 2. Geothermal Electric FITs in Non-EU Countries

Country	FIT Payment Structure*	Requirements for FIT Payment	FIT Payment Duration (Years)	Geothermal FIT Payment (€/kWh)	Geothermal FIT Payment (US$/kWh)	Geothermal Cumulative Installed Capacity (MW) (2010)	Cumulative Geothermal Capacity Supported by FIT (MW) (as of 2010)
Croatia	Fixed	<1 MW	12	0.18	0.25	0	0
		>1 MW		0.17	0.24		
Ecuador	Fixed	Mainland	15	0.093	0.13	0	0
		Islands (Galapagos)		0.102	0.14		
Serbia	Fixed	N/A	12	0.075	0.11	0	0
Switzerland	Fixed	<5 MW	20	0.34	0.48	0	0
		<10 MW		0.3	0.42		
		<20 MW		0.23	0.32		
		>20 MW		0.19	0.27		
Taiwan	Fixed	N/A	20	0.12	0.17	3.3[33]	0
Turkey	Fixed	Base rate	10	0.07	0.10	81	0
		Bonus: Steam or gas turbine domestic content		0.009	0.01		
		Bonus: Generator and power electronics domestic content		0.05	0.07		
		Bonus: Steam injector or gas compressor domestic content		0.05	0.07		
Uganda	Fixed	N/A	20	0.05	0.07	0	0

Sources: (Bertani 2005; Bertani 2010; Electricity Regulatory Authority 2010; Gipe 2011; Holm et al. 2010; IGA 2011; Kolarevic 2009).

There are several likely reasons for the lack of market response to geothermal FITs thus far. First, many of the countries in Europe have comparatively low levels of hydrothermal resources and are attempting to support EGS, which is an emerging technology (not fully commercial) and therefore has proven expensive and challenging to develop to date (GEOFAR 2010). Second, several of the geothermal FITs are comparatively new (or have been recently amended); geothermal developers have not yet had time to react given the long development lead times required of geothermal projects. As a result, policymakers have not had the benefit of calibrating their policies according to established policy track records (as they have been able to do with wind and solar FITs). Third, the payments offered under the FITs may be insufficient to support geothermal development because the payments are too low to compensate for the price of the geothermal resource or because of other country-specific risks (e.g., political risk or currency risk) (GEOFAR 2011; Goldbrunner 2010). Fourth, the presence of other technical and non- technical barriers to geothermal development (e.g., unmitigated exploration risks or lack of transmission) may serve as a barrier despite the presence of otherwise attractive payment levels (GEOFAR 2009b; Hurlbut 2012). Finally, some of the FITs are available only to small projects whereas geothermal project development has focused on larger-scale projects (e.g., Italy).

A key take-away from the international survey is that the presence of FITs alone has not driven results comparable to those achieved for wind and solar power under European FITs or to those achieved for geothermal under PURPA in the United States. On the other hand, international experience demonstrates that many countries are experimenting with crafting policies that are more specifically tailored to geothermal than North American policies (e.g., RPS and PURPA) have been in the past. Geothermal FIT policies are continuing to evolve and mature and there remains room for innovation and further research into international designs. FITs can be designed to take country-specific geothermal resources and risks into account and may be part of a broader package of geothermal risk mitigation policies, including exploration support or expedited permitting (Miethling 2011).

Unlike many of the countries in the international survey, the United States has a strongly identified and unexploited hydrothermal resource as well as a significant amount of as-yet unidentified hydrothermal resource. It is reasonable to assume, especially given the experience with PURPA, that appropriately crafted and targeted FITs in the United States could have a greater impact on market growth than in other parts of the world.

3.3.3. Approaches to Setting FIT Payment Levels

A primary step to establishing targeted FITs for geothermal is determining the payment levels that will be offered to generators. As explored in a recent NREL report, a broad range of approaches can be used to set payments for FITs or other performance-based incentives (Gifford et al. 2011). The key considerations when developing FIT payment levels include:

- Basis for setting payment levels. Payment levels can either be set based on the generation cost of a certain technology (a "cost based" approach) or pegged to measures of value such as the avoided cost of energy and/or avoided externalities.
- Contract length. As shown in Tables 1 and 2, the contract length varies widely between FITs. For generation cost based payments, the contract length can have an impact on the $/kWh payment to the generator. Shorter contracts require a higher payment than longer-term contracts to achieve the same rate of return.
- Model for setting payment levels. Different jurisdictions utilize different spreadsheet models to set their payment levels. The key policy choices related to model design are whether to employ cash flow forecasts or recovery factor analyses and whether to utilize a pre-tax or after-tax analysis. In a separate effort, NREL surveyed international FIT practices and U.S. policy requirements, and developed an after-tax discounted cash flow model called CREST[38] that includes a geothermal model (Gifford et al. 2011).
- Type of input data. Models for setting payment levels can range from high-level and comparatively simple to highly complex models. The number of inputs and the granularity of those inputs can to a great extent determine model complexity. For example, modelers can choose to use a single input for "installed cost" or they can choose to break installed costs into labor cost, detailed equipment costs, and other upfront (e.g., development, permitting, financing-related) costs.
- Input selection. A range of values that can be considered "reasonable" for each input. Policymakers have a choice as to whether they select values at the upper or lower ends of these ranges. To a great extent, these choices will determine how aggressive or conservative[39] the ultimate FIT payment level is and thus the volume of new generation that will be able to participate under the policy.

As seen in Tables 1 and 2, the manner in which policymakers choose the basis for geothermal payment level setting and select modeling inputs can result in widely different FIT payment levels. Given the magnitude of the costs and risks of exploration and development, a key choice for U.S. policymakers will be whether and how these costs are incorporated into the development model.

The two primary approaches to setting FIT payment levels are to base the payment on a measure of value (e.g., avoided cost) or to base the payment on expected generation cost of eligible projects (Grace et al. 2008). In North America, the current state FIT in California and the municipal FIT in Sacramento are based on avoided cost (plus adders reflecting externalities), whereas the remainder are generation cost-based policies. Cost-based FITs dominate world FIT policy methodology. This report focuses on generation cost-based FITs in order to highlight the wide number and variety of policy choices that could have implications for geothermal projects. The development of value-based payments could be the focus of future research.

This section reviews the geothermal policymaking and approach to setting payment levels in two countries, Germany and Spain, in order to explore this and other issues in greater detail.

3.3.3.1. Geothermal FIT Policy and Payment Level Setting in Germany

Germany's fleet of installed geothermal capacity remains comparatively small. The first system was installed in 2003 and there were five systems totaling 7.5 MW by the end of 2010 (BMU, 2011a). Compared to the U.S. generation fleet, these systems are very small scale and embrace less common approaches such as Kalina cycles, EGS, and combined heat and power systems (GEOFAR 2010).

Germany's renewable energy FIT has been in place since 1991. The first tariff for geothermal power was introduced in 2000 with the passage of the Renewable Energy Law: the Erneuerbare- Energien-Gesetz (EEG).[40] The initial 20-year tariff level was set at 8.95 €cent/kWh (12.53 ¢/kWh) for systems up to 20 MW in size and 7.16 €cent/kWh (10.02 ¢/kWh) for systems larger than 20 MW. The EEG included annual decreases in the available FIT payment for most generators according to a fixed schedule (i.e., a degression schedule[41]). Geothermal payments were not subject to degression, however, in order to account for long development lead-times. The first geothermal power plant in Germany was built under the FIT in 2003.

The German Ministry of Economy and Finance (BMWI) and the Ministry for the Environment (BMU)[42] co-financed the first geothermal electricity plant in order to gain a better understanding of geothermal economics.[43] Based on the experiences with the first power plant and additional research, policymakers concluded that the tariff levels under the EEG were too low to support geothermal market expansion, especially for small-scale generators (Jacobs 2012). In the 2003 FIT progress report, the BMU recommended adjusting the tariff to better account for geothermal plant economies of scale (BMU 2003). The revised EEG of 2004 introduced two additional size categories below 20 MW, according to the table below:

The 2004 law also introduced a new degression of 1% annually, but delayed the start of the decrease until 2008 to account for the lengthy development timeline.

In 2007, the BMU again released a progress report that evaluated the adequacy of German FIT payment levels. The report concluded that an additional increase in the tariff levels would be necessary in order to account for higher actual drilling costs resulting from increased activity in the oil and gas sector (BMU 2007). The revised 2008 EEG introduced simplified size categories with higher payment levels as can be seen in the table below; one goal was to provide revenue to cover the assumed project drilling costs. The law also included bonus payments for systems that went into operation before 2016, that used combined heat and power, and that used EGS technology. These bonuses are additive such that an EGS CHP system that is under 10 MW and installed prior to 2016 would receive €0.27/kWh (0.38 $/kWh) for 20 years.

Table 3. Size Categories and Tariff Payment Levels under Revised German EEG (2004)

Geothermal plant size	€/kWh	$/kWh
< 5 MW	0.15	0.21
5 MW – 10 MW	0.14	0.20
10 MW – 20 MW	0.0895	0.13
> 20 MW	0.0716	0.10

Table 4. Simplified Categories and Tariff Payment Levels for 2008 German EEG Revision

Geothermal plant size	€/kWh	$/kWh
< 10 MW	0.16	0.22
> 10 MW	0.105	0.15
Bonus: Installed prior to 2016	0.04	0.06
Bonus: Combined heat and power	0.03	0.04
Bonus: EGS	0.04	0.06

On January 1, 2012, the German FIT was again revised to include a single payment level for geothermal projects of €0.25/kWh (0.35 $/kWh) for 20 years. EGS plants are eligible for a €0.04/kWh (0.056 $/kWh) bonus payment. The rates from 2007 were consolidated into one higher rate, and the new rate reflects that drilling costs had been more uncertain than anticipated[44] and that financing costs were higher than previously modeled because of larger project risks. Degression is set at 5% a year but does not begin until 2018. This latest version of the German geothermal FIT payment values are captured in Table 1.

The key parameters used to calculate the German FIT payment level in 2011 are contained in Table 5 below. It is noteworthy that drilling costs are assumed to make up the largest share of the installed cost for geothermal plants in Germany. These drilling costs are based on experience from six geothermal projects with 11 production wells between them (total 47,000 meters of drilling) (Jacobs 2012). For each project, drilling costs amounted to between €20 and €28 million ($28-$39 million).

Table 5. Key Parameters for Calculation of German FIT

Debt to equity ratio	59/41
Return on equity	12.3%
Interest rate	6%
Weighted average cost of capital	9.3%
Contract length	20 years
Installed costs	10,900 € per kW installed (15,260 $/kW)
Inflation rate	2% per year
Drilling costs (% installed costs)	47
Power plant costs (% installed cost)	27
Development costs (% installed cost)	6
Labor	100,000 € per year ($140,000 per year)
Full load hours per year	8,000

Source: BMU (2011a).

It should be noted that the weighted average cost of capital of 9.3% reflects costs of capital for the exploration, power plant construction, and operation phases. The German government analysis assumed that the exploration phase is 100% equity financed, with a 20% return on equity (ROE) (BMU 2011a). At the end of the development process, it is assumed that the project will secure permanent financing with a debt to equity (D/E) ratio of 70/30, an 11% ROE and a 6% interest rate. The weighted average return on equity from exploration to construction is 12.3% and the weighted average D/E ratio is 59/41.

3.3.3.2. Geothermal FIT Policy and Payment Level Setting in Spain

Spain's FIT has gained international attention because it has supported a strong and sustained wind energy market growth (GWEC 2011) and because it created an unprecedented (and short- lived) solar photovoltaic market boom in 2008 (Ciesielska et al. 2011).

Most of the Spanish FIT payment levels are based on the generation costs of each specific technology. Geothermal, however, does not receive a cost-based payment and instead receives a general payment available to "emerging" technologies such as wave power, ocean thermal energy conversion technologies, and tidal power (Jacobs 2012).

The payment available to geothermal power plants in Spain has remained relatively stable over the years. The initial emerging technology tariff of 6.73 €cent/kWh (9.42 ¢/kWh) was enacted into law in 1999. From 2004-2007, the tariff was set at 90% of the average retail electricity rate, which was 6.49 €cent/kWh (9.09 ¢/kWh) in 2004 and 6.6 €cent/kWh (9.24 ¢/kWh) in 2005 and 2006. In the 2007 amendment to the FIT law, the fixed emerging technology tariff payment level was set at 6.89 €cent/kWh (9.65 ¢/kWh) (Jacobs 2012). The available payment level, which is indexed to inflation, has since increased to 7.441 €cent/kWh (10.42 ¢/kWh) in 2011.

The FIT payment level is available for the first 20 years of operation. After the first 20 years, geothermal generators are eligible for a slightly lower payment level that is available for the remainder of their operational life. Currently, the payment level available after the 20th year is 7.0306 €cent/kWh (9.84 ¢/kWh).

No geothermal power has been developed to date in Spain (Sanchez-Guzman and Garcia-de-la- Noceda 2010). This is primarily because Spain does not have strong hydrothermal resources on the mainland[45] and likely because the current emerging technology FIT payment level is not high enough to support hydrothermal or EGS development.

3.3.3.3. Lessons from Germany and Spain

There are several key lessons from experiences in Germany and Spain.

- Policy objectives matter. The manner in which policy objectives are interpreted through policy design can have a dramatic impact on the payment levels that generators receive. Germany's decision to provide a bonus payment for EGS (separately from hydrothermal), for example, resulted in much higher payment levels than Spain's technology neutral payment level for emerging technologies.
- Exploration risk can be considered. The payment level setting process can be used to reflect exploration and development risks. Setting the payment level based on generation cost enables policymakers to specify the degree to which exploration risk is assumed to be a project cost driver. In Germany, for example, these risks are reflected in the significant share of project costs that derive from the development and exploration phases and in the financial assumptions. Incorporating a greater degree of exploration and development risk into the payment level results in higher payment levels that can allow a broader range of potential projects to be developed (but also creates higher policy costs if there is a significant volume response).
- Generic FIT payments may not be effective. The experience in Spain demonstrates that generic FIT payments not set based on generation cost may be too low to support geothermal development, particularly in countries without a strong hydrothermal resource.

4. FIT Design to Support Geothermal Development in the United States

4.1. Integrating FITs with Geothermal Targets

FITs can be designed specifically to support geothermal energy development in the United States, by considering FIT design issues that are specifically relevant to the unique characteristics of geothermal power plant development and operations in the U.S. context.

Many FITs are directly linked to national or state energy targets. The 2008 FIT law in Germany, for example, explicitly specifies that the intent of the FIT is to meet the national renewable electricity target of at least 30% by the year

2020. FITs can also be linked explicitly to targets for individual technologies. The linkage of FIT policies with mandatory renewable energy targets gives investors a better sense of the policies' horizons and increases investor security (Deutsche Bank Climate Change Advisors 2011). If geothermal FITs are established, they can be explicitly linked with mandatory geothermal targets through law or regulation. While there is not a national U.S. renewable electricity target, there are currently 29 states that have mandatory RPSs and 8 states that have voluntary goals. FITs are one of the potential mechanisms that can be utilized to achieve RPS targets and objectives (Cory et al. 2009). Fourteen states have established specific targets for solar electricity and several have established specific goals for other technologies or applications (e.g., wind, distributed generation, etc.). At present, however, opportunities to integrate geothermal FITs with existing RPS targets are limited since no targets specific to geothermal energy exist.[46]

Policy design assessment for lowering investor risk. Setting mandatory and specific targets for geothermal can improve investor confidence in a state's commitment to geothermal procurement. Setting specific mandates for geothermal can also form the basis for states to set geothermal- specific payment levels under PURPA and reduce concerns about states' ability to set premium wholesale rates.[47]

4.2. Technology Eligibility and Differentiation

FITs can consider technologies in two distinct ways. Eligibility refers to whether a certain technology, ownership type, or project size is allowed to participate in (or is excluded from) a given policy, whereas technology differentiation refers to how FIT payment levels are targeted to certain technologies, sizes, ownership models, or locations. As can be seen in Tables 1 and 2, international FITs generally have few eligibility requirements outside of overall project size caps, but are differentiated in ways that reflect diverse policy objectives. U.S. policymakers would be able to customize FITs to support different types of geothermal installations in a number of different ways. Based on recent U.S. development trends, FIT differentiation strategies could include:

- Geothermal resource type. In order to support the different types of geothermal resources that are currently being explored, including hydrothermal, EGS, co-produced and geopressured, FIT payment levels could be tailored and differentiated to have a different payment level for each resource type. Such an approach acknowledges that each of these applications has different cost structures and would ensure that each receives adequate revenue for project development, but not more than is required (e.g., payments appropriate for EGS might result in overpayment for other applications).

- Project size. A broad range of geothermal project sizes are currently under development in the United States, with a significant percentage of proposed plants smaller than 50 MW (Jennejohn 2011). Since economies of scale can exist as projects get larger, FIT payments could be differentiated by size in addition to technology. As shown in Table 2, Switzerland uses four project sizes: <5 MW, 5-10 MW, 10-20 MW and 20+ MW.[48]
- Resource vintage. Geothermal development is proceeding both at existing and new geothermal fields. The risks and costs of identifying and developing a greenfield geothermal site are markedly different from those involved in expanding an existing site or building on an identified reservoir. Tariff differentiation could take this difference into account as well.

Policy design assessment for lowering investor risk. The choice of eligibility or differentiation does not have direct impact on project risk, although it will impact which projects are more attractive to investors. The primary concern from an investor perspective is whether the payment available is sufficient to cover operating expenses and meet the required returns for the project.[49]

4.3. Incorporating Exploration Expense in FIT Payment Levels

Setting FIT payments often includes assumptions about the cost of measuring the available resource (e.g., wind speed, solar insolation, and wood fuel availability). For geothermal, establishing resource availability and adequacy requires going beyond desktop study or relatively inexpensive meteorological measurements, requiring costly field-based study, which often includes significant exploration and test well drilling activity. These resource assessment activities can comprise a much greater share of total project cost and a greater portion of the overall development timeline than for other renewable energy technologies. For geothermal, there is also a greater risk that these initial resource exploration activities will not produce a viable project. Multiple locations may need to be explored before a specific project location enters the design and permitting process. As a result, the manner in which estimates of exploration expense are incorporated into FIT payment setting models is a serious consideration. Options for incorporating exploration expense include:

- Include estimated total exploration costs. Incorporate an assumed exploration success rate, investor return requirement, and commensurate level of exploration expense explicitly in FIT payment setting. Using this method, the developer's exploration risk is reflected through a higher FIT payment level. Different assumed exploration success rates and investor return requirements can be selected to establish different FIT payments for regions with existing geothermal facilities compared to those which have yet to be proven. The modeled exploration costs and return requirements are a policy choice and can reflect a preference for more aggressive or conservative payment setting.
- Include a fraction of estimated exploration costs. Attributing only a portion of actual exploration expenses to a project by assuming developers will cost-share exploration costs, exploration costs could be spread over several power plants, or that a portion of the costs could be decreased by other public-sector cost sharing.[50]
- Ignore exploration costs. No exploration costs assumed in payment setting.

Policy design assessment for lowering investor risk. Acknowledging and accounting for exploration costs in FIT payment setting is one of the most effective ways to reduce investor risk and encourage the flow of capital, from both existing and new sources, into geothermal project development. Assurance of a financeable contract that enables the recovery of reasonably incurred exploration expenses provides a signal to the market to invest, notwithstanding the fact that not every exploration is expected to produce a viable thermal resource.

4.4. Standard Pricing and Standardized Off-Take Contracts

While competitive solicitations play a critically important role in the renewable energy market place, their cyclical and often fast-moving announcement, bidding, and negotiation schedule favors renewable energy technologies with shorter development lead times than geothermal projects. Developing a long-term, standard offer price and standard contract for specified eligible facilities is an effective tool for supporting long lead-time projects. Standard offer prices and contracts can reduce the expense associated with periodic solicitations by removing the bid security, legal, bid preparation, and other transaction costs of pursuing a competitive PPA.

Policy design assessment for lowering investor risk. FITs provide generators with standardized prices and contracts that can reduce the contract price risk, transaction costs, and uncertainty of PPA negotiation. This has the potential to reduce investor return requirements and increase the pool of available capital.

4.5. Payment Structure

As shown in Table 1, FITs can be structured as fixed "all-in" prices (that include the power and the attributes, or renewable energy certificates) or premium payments that float on top of (and are additive to) wholesale market prices.[51] In general, premium payment options are utilized to encourage generators to participate in wholesale electricity markets in order to support deeper and broader competition, and to encourage generators to respond to market signals. Geothermal power plants are baseload generators and unlikely to adjust their behaviors based on market prices. Moreover, several electricity markets in the United States do not have competitive wholesale markets with liquid spot prices. As a result, premium payments may not be a practical option for geothermal FITs in some locations.

Policy design assessment for lowering investor risk. From the perspective of investors, fixed prices[52] are lower risk than premium payments that leave greater uncertainty because the total payment level varies with market prices.

4.6. Contract Duration

Geothermal plant life is typically assumed to be at least 30 years (Lovekin and Pletka 2009; PT Castlerock Consulting 2010). As can be seen in Table 1, international FITs are paid out over 10-20 years, with the exception of Spain which guarantees a payment for the entirety of project life (albeit at a lower payment level after year 20). As shown in Figure 6 in Section 5, under a cost-based FIT, a longer contract length would lower the $/kWh payment paid to generators since project costs would be spread out over a longer period of time. A longer-term contract would therefore create greater opportunity for geothermal contracts to serve as an effective hedge against fossil fuel prices (or even create immediate savings if the avoided cost rises above the tariff payment).

Longer contract durations also provide increased revenue certainty for project investors. This is likely to increase the number of investors willing to finance geothermal projects and reduce the cost of capital. Reductions in the cost of capital would stem from both the increase in the number of active investors and the reduction in risk provided by the long-term, stable cash flows.

Policy design assessment for lowering investor risk. To the extent that post-contract market revenue projections are utilized in setting a FIT payment,[53] longer contract durations reduce investor risk by reducing the proportion of lifetime revenues exposed to market price risk. If, however, FIT payments are designed to attain target investor returns prior to contract termination, investors should be indifferent to contract duration.[54]

4.7. Policy and Contract Timing

Core characteristics of geothermal development, particularly long development horizons and the presence of substantial exploration risk, make aspects of policy timing particularly important. To have its greatest benefit in mitigating developer risk and therefore reducing financing costs, a FIT policy would (1) be of sufficiently long duration to be available upon project commercial operation to a developer investing in exploration-phase activities and (2) set and hold a FIT price available during the operation of a geothermal project well in advance so a geothermal developer could make exploration and development investment decisions based on specific and reliable revenue expectations. Such a FIT policy could reduce or eliminate the need for developers to build in a risk premium due to competitive RFP processes.

Establishing FIT cost-containment mechanisms, such as volume caps, price adjustment mechanisms, and/or automatic triggers for policy review, appear to help avoid FIT oversubscription and provide greater certainty about policy outcomes (Kreycik et al. 2011). It is worth noting that the experiences with oversubscription have been primarily with FITs established for solar and wind power, technologies with substantially shorter lead times and greater rates of technological advance (price evolution) than geothermal. For this reason, the potential for an explosive project development response to a geothermal FIT would appear far lower than for solar and wind; policymakers should have adequate time to adjust. Nonetheless, cost containment and policy review mechanisms appear to be a standard practice for FIT implementation today. If caps are implemented for geothermal, it is useful to acknowledge the policy tradeoffs. On the one hand, implementing volume caps and typical price adjustment mechanisms can undermine the benefit of assured market access and can increase developer risk. On the other hand, holding a price and access open for as long as 4- 7 years in advance (i.e., the development horizon for a geothermal plant) could restrict the establishment of meaningful cost control practices.

Policy design assessment for lowering investor risk. At a minimum, the presence of a stable FIT policy that could be relied on by an early-stage geothermal developer investing in exploration- phase activities could decrease developer risk. A policy designed to balance investor risk mitigation with policymaker/regulator needs to minimize ratepayer risk might have some or all of the following characteristics:

- Long duration. Be established for a sufficiently long duration to span the period for securing a long-term revenue stream from exploration decision to commercial operation.
- Developer queue with capacity cap. If a capacity cap is established for a geothermal FIT, it may be accompanied by a developer queue to provide regulators with a means to track development progress in response to the FIT while providing developers in the queue with the ability to rely on access to the FIT. This might include developer registration for a specific number of MW, accompanied by a modest $/MW deposit to maintain a queue position. First registrants in such a queue might receive a longer assured access to known price than

developers later in the queue. To avoid speculative queuing and allow for removal of developers unprepared to move forward, an increasing deposit might be required over time to maintain a preferred queue position[55].

- Adequate lead time for policy adjustments. Allow regulators to establish or modify quantity/cost caps and tariff prices with a lead-time that balances the needs of policymakers to contain costs and the value to developers of assured access to a FIT at known price. For projects whose remaining development activities are expected to extend beyond this lead-time, their investors will need to understand that caps might be reached (if applicable) and that adjustments might be made to the nonetheless cost-based tariff. The lead-time could be fixed (e.g., 4 years), or might be adjusted within a range based on the MW in the queue.

4.8. Payment Indexing over Time

In some jurisdictions, the FIT payment (or a portion thereof) is adjusted annually to account for inflation. Since renewable generators such as solar and wind have small operation and maintenance (O&M) expenses, some analysts have argued that they do not require annual inflation adjustments (Boonin 2008). Geothermal plants, however, can have proportionately higher O&M expenses. In addition to power plant maintenance, geothermal power plants also require well field maintenance. Geothermal field production, for example, will decline over time depending on factors such as plant capacity and the rate of natural reservoir replenishment. In order to compensate for geothermal resource declines, power plants drill additional "make-up" wells to boost production. It has been estimated that annual make-up drilling costs correspond to 5% of initial drilling cost (Hance 2005). During the past several years, the cost of drilling has been particularly volatile, rising up to 64% in the United States between 2004 and 2008 and then subsiding by 30% between 2008 and 2010 (Augustine et al. 2010; Lovekin and Pletka 2009). Given both the significance of make-up drilling costs and the volatility, an alternative to using a general inflation rate as an index for FIT payment adjustment is to index geothermal payments to follow the cost of drilling over time, provided that a representative, independent, and visible drilling cost index is available.

Policy design assessment for lowering investor risk. Given the potential size of geothermal maintenance O&M expenses, indexing a portion of a FIT payment to inflation, drilling costs, and/or other relevant indices would reduce project risk and assure investors that long-term O&M costs were considered.

4.9. Adjusting the Policy

There are a range of different approaches to adjusting FIT policies over time in order to reflect changing market conditions and policymaker objectives. Adjustments can be characterized by what triggers the adjustment and the type of adjustment that is made. Triggers can include milestones such as the installation of a certain amount of capacity or the passage of a specified amount of time. The type of adjustments that can be made include an automatic adjustment to the available payment level, a hard cap on the policy after which no new FIT contracts are available, or the instigation of a full policy review (Deutsche Bank Climate Change Advisors 2011). In Germany, for example, an automatic price decline is triggered every year, whereas a full policy review is triggered every four years (BMU 2011b). Policy adjustments are a key concern for technologies such as solar photovoltaics (PV) that have short development timelines, an almost universally available resource, and the ability to scale-up rapidly. Since geothermal development timelines are long, however, time-based policy adjustments may not be necessary and instead the policy can link adjustments to the achievement of capacity targets or make them contingent upon the outcomes of periodic review.

Policy design assessment for lowering investor risk. An ideal policy would have infrequent adjustments made on a schedule that is known in advance and would utilize a process that is as transparent as possible.

4.10. Setting Geothermal FIT Payment Levels in the United States

Depending on a given state's policy objectives, its geothermal resource, and other factors such as barriers, risks, and existing policies, actual FIT designs could vary widely. We considered the ability of a policy design element to lower investor risk as a simple analytical metric to distinguish between different design options. The evaluation criteria used to choose between different FIT designs will also vary from state to state and may give weight to stakeholder groups beyond investors (e.g., ratepayers). Moreover, this section does not attempt to compare or contrast FIT designs with other policy types.

Table 6 summarizes several of our findings.

Table 6. Key Findings Regarding FIT Policy Design in the United States

Policy Design Issue	Design Option Potential for Lowering Investor Risk
Integrating FITs with Geothermal Targets	Setting mandatory and specific targets for geothermal can improve investor confidence in a state's commitment to geothermal procurement. Setting specific mandates for geothermal can also form the basis for states to set geothermal- specific payments under PURPA and reduce concerns about states' ability to set premium wholesale rates.
Technology Eligibility and Differentiation	The choice of eligibility or differentiation does not have direct impact on project risk. What matters from an investor risk perspective is whether the FIT payment available to geothermal generators of various sizes and types is sufficient to cover operating expenses and meet investors' required returns.[56]
Incorporating Exploration Expense into FIT Payments	Accounting for exploration costs when setting the FIT payment (i.e., by building exploration cost assumptions into the price setting model) reduce investor risk and encourage the flow of capital into geothermal development. Assurance of a financeable contract that enables the recovery of some level of estimated exploration expenses provides a signal to the market to invest, notwithstanding the fact that not every exploration is expected to produce a viable thermal resource.
Standard Pricing and Off-Take Contracts	FITs provide generators with standardized off-take contracts that can reduce power project contract price risk and transaction costs. This has the potential to reduce investor return requirements and increase the pool of available capital.
Payment Structure	Fixed FIT prices are lower risk (from the investors' perspective) than premium FIT payments, reducing the uncertainty associated with exposure to fluctuating market electricity prices.

Table 6. (Continued)

Policy Design Issue	Design Option Potential for Lowering Investor Risk
Contract Duration	If post-contract market revenue projections are utilized in setting a FIT payment, longer contract durations reduce investor risk by reducing the proportion of lifetime revenues exposed to market price risk.
Policy and Contract Timing	The presence of a stable FIT policy that could be relied on for a period of several years by an early-stage, exploration-phase geothermal investor could decrease overall risk, thus reducing/eliminating developer risk premiums for RFP bids.
Payment Indexing Over Time	Given the potential size of geothermal operations and maintenance expenses, indexing to inflation, drilling costs, and/or other relevant indices would reduce the risk that project revenues will be adequate.
Adjusting the Policy	Adjustments that are made infrequently and on a schedule that is known in advance and using a process that is as transparent as possible increases investor confidence. Geothermal policies may not need to be adjusted as frequently as technologies with rapidly declining costs such as solar PV, or with shorter development timelines like solar PV and wind.

5. Analyzing the Impact of Policy Design Choices on the Levelized Cost of Energy from Geothermal Electricity Generators

5.1. Using LCOE to Inform FIT Payment Level-Setting

FITs are typically set using either a constant price over the term of the contract, or a fully- or partially-escalating payment. This analysis uses the nominal LCOE as the evaluation metric for illustrating the quantitative impact of FIT policy design choices.[57] The nominal LCOE is the constant payment, expressed in nominal cents per kWh, which is sufficient to cover all of the costs of building and operating a power plant over its economic life, including meeting the minimum return requirements of investors. An LCOE has the equivalent economic effect over the analysis term (i.e., same net present value) as a time-varying payment stream covering the total cost of building and operating a generating plant over its life. This section takes a cost- based approach to FIT payment-setting and calculates a long-term contract value,[58] which will enable the modeled geothermal project to secure financing by both covering its operating expenses and meeting investors' minimum return requirements. For these purposes, the calculated LCOE over the duration of the FIT is equal to the assumed FIT payment.

While assuming that the calculated LCOE is equivalent to the FIT payment can provide valuable insights for the purposes of this report, the cost of geothermal projects will vary based on a range of factors including project technology, size, and resource quality. Furthermore, the broader policymaking context may include additional objectives, such as rapid deployment, diversity in project size or location, or least cost deployment, which may result in selecting a FIT payment that is more aggressive or more conservative[59], respectively (as noted in Section 3.3.3). For these reasons, a single calculated LCOE model result can be used directly as the FIT payment, but policymakers may wish to use analyses such as this one to inform a FIT payment-setting process taking into account both cost variation and policy objectives. With this caveat, the terms LCOE and FIT payment (and references to ranges of these values) are used interchangeably for the remainder section.

5.2. Objectives: Understanding the Potential Impact of FIT Design Choices

A FIT represents not only the availability of a long-term contract, but also a potential shift in investor confidence with respect to geothermal investment. While the technical risk of identifying a viable resource remains (i.e. exploration risk), a FIT or similar policy alleviates the risk of whether a viable contract will be available to provide a means to sell the output and attract financing.

To provide a simplified illustration: if the probability of successful exploration for an economically-viable geothermal plant in a specified region is thought to be 33%, and the developer believes it has a 1-in-3 chance of securing a long-term power contract, then the developer's probability of recovering its exploration investment and securing permanent financing to construct and operate a power plant is 33% * 33% = 11%. With a FIT available to guarantee access to a cost-based revenue stream, the risk of contracting and permanent financing is removed (i.e., 100%) and the odds of the developer's recovery of exploration capital increase to 33%, the probability of successfully identifying a viable resource. A simple understanding of risk and reward tells us all we need to know: the investment with an 11% probability of success will need to have a higher return potential than the investment with a 33% probability. The higher risk environment will also make it more difficult to attract the capital necessary to fund development activities. The cost of investment capital available to a geothermal project will behave similarly: the cost of capital decreases as the probability of success increases. Similarly, the availability of capital will increase as probability of success increases.

Competitive procurement for long-term contracts can mitigate only some of this risk, providing the possibility of access to a long-term fixed revenue stream with a credit-worthy entity but without the assurance of either the payment level or access to the revenue (both dictated by whether the project wins the competitive processes). Where binding policy support is not in place over a horizon long enough to earn a commercial return, or where long-term contracts are not offered, there is even less assurance of access to a revenue stream sufficient to attract financing.

It is difficult to precisely project the degree to which financing costs might be reduced. Using a range of representative geothermal cost data and 'what if' analysis, the potential impact on LCOE of reducing financing risk to geothermal developers was explored. It is intended to meet several objectives for policymakers with respect to LCOE analysis and FIT payment-setting, including:

- To identify the range of LCOEs that result from different assumptions about the cost, duration, and financing terms of geothermal development;
- To use these LCOE results to consider how to craft geothermal (FIT) policy that brings the greatest reduction in geothermal LCOE;
- To illustrate how a geothermal FIT can increase the availability of and/or reduce the cost of investment capital necessary to finance the exploration, development, and construction of geothermal facilities.

These objectives are accomplished herein by identifying a range of geothermal cost, financing, and operational input assumptions and conducting sensitivity analyses to establish a continuum of LCOE results for representative geothermal projects. Varying key inputs such as development duration and financing terms demonstrate potential market responses to different FIT designs.

The resulting range of LCOE values is a proxy for the potential continuum of FIT payments that might be considered in particular circumstances and locations. Examining the sensitivity analysis can help policymakers identify FIT designs and other incentives that help minimize geothermal LCOE while making investment attractive.

5.3. Methodology: Literature Review and CREST Modeling

5.3.1. Literature Review

A range of geothermal cost, performance, and financing inputs was developed through a detailed literature review. Publicly available geothermal cost research was compared and used to develop the series of inputs for LCOE modeling. A summary of the literature review is provided in Appendix A.

5.3.2. The CREST Model

Geothermal LCOEs were subsequently calculated using the CREST model. NREL developed CREST—an after-tax discounted cash flow model—based on a survey of international modeling practices and U.S. policy requirements (details captured in Gifford et al. 2011). A geothermal- specific version was released in 2011 and was used for this analysis.

CREST is an economic cash flow model designed to enable PUCs and the renewable energy community assess projects, design cost-based incentives (e.g., feed-in tariffs), and evaluate the impact of tax incentives or other support structures. CREST is a suite of three analytic tools, for solar (photovoltaic and solar thermal), wind, and geothermal technologies, respectively and it can be downloaded at https://financere.nrel.gov/finance/content/crest-cost-energy-models. The CREST models were designed for use by state policy makers, regulators, utilities, beginning developers or investors, and other stakeholders. Inputs include the variables under study in the sensitivities and scenario analysis (e.g. installed costs, O&M costs, financing assumptions, contract duration, etc.). The models allow users to:

- Estimate the Year One cost of energy (COE) and levelized cost of energy (LCOE) from a range of solar, wind and geothermal electricity generation projects
- Inform the process of setting of cost-based incentive rates
- Gain understanding of the economic drivers of renewable energy projects, which lead to the calculated COE and LCOE

- Understand the relative economics of generation projects with differing characteristics, such as project size, resource quality, location (e.g. near or far from transmission) or ownership (e.g. public or private).

5.3.3. Enhancements to Geothermal CREST

The co-authors of this report were also the architects of the CREST model. As part of the scope of this report, NREL directed several enhancements to the geothermal CREST model to facilitate the analysis contained herein. Specifically, adjustments were made so that the capital structure (combination of debt and equity) and cost of invested capital (equity return requirement and debt interest rate, if applicable) can be specified separately for the exploration drilling,[60] confirmation drilling, site construction, and permanent financing phases.

Because CREST allows model users to input different financing assumptions for each phase of geothermal development, policymakers can test the sensitivity of geothermal LCOE to the availability and cost of capital.

5.3.4. Modeling the Phases of Geothermal Development and Financing

The development of a geothermal power plant includes at least four distinct activities, modeled in this analysis over three phases of development. For cost estimating purposes, and based on the literature review, this study views geothermal development as including the following four activities, each with its own financing assumptions:

- Exploration – desk-top studies, initial surface exploration, temperature gradient drilling, and, in some cases, deep exploratory drilling[61]
- Confirmation – continued deep exploratory drilling, resource assessment, and initial production wells
- Well Field Construction – completion of well field drilling, including remaining production wells and injection wells
- Power Plant Construction – including interconnection.

The manner in which geothermal development is financed influences LCOE and FIT payment- setting. Geothermal projects typically involve different financing sources at different phases of project development (Deloitte 2008; Richter 2009; Salmon et al. 2011). These sources come in different ratios (e.g., debt to equity) by phase, and have different risk profiles and terms (e.g., return requirements, interest rates, and repayment schedules). For the purpose of this analysis, the CREST model analyzes geothermal development as having three financing phases, each with its own cost and financing assumptions. Equity investments are assumed to fund both the exploration and confirmation drilling phases. While some forms of debt (including so-called mezzanine debt, which has characteristics of both debt and equity) may have once been available for confirmation drilling and related activities, this analysis assumes that debt is first introduced at the well field and power plant construction phase once the availability and adequacy of the geothermal resources has been entirely proven, and once revenue expectations have been established. Well field and power plant construction are assumed to be financed at the same time, on the same terms, and occur more or less in parallel.

5.4. Inputs for Calculating Geothermal LCOE

As described in Section 3, a key component of the FIT payment-setting process is determining the range and complexity of modeling inputs.[62] The CREST model used for this analysis allows for an appropriately-detailed set of installed cost, performance, incentive, and financing assumptions. Each modeling input will have a continuum of possible values. Selecting which points on the continuum are used for modeling can be tied to policymaker preferences for more aggressive or conservative tariffs.

It is important to understand that the inputs and associated LCOEs are intended to be illustrative rather than representative of the cost of a specific operating or proposed geothermal project. The most important function of this analysis is to explore the relative change in LCOE caused by changes in each of the factors explored in the sensitivity analyses, and not the absolute value of the calculated LCOE itself. A range of assumptions has been used for this report that, based on the literature review, should capture the majority of project costs and characteristics.

5.4.1. Exploration Success, Confirmation Duration, and Project Cost Inputs

As described in Section 4.3, incorporating expected exploration expenses in the FIT payment- setting process is critical to establishing investor confidence and spurring investment. The success of geothermal exploration can vary widely, including the difference between exploring for the expansion of production in an area where geothermal resources have already been found and exploring in unproven territories. It is assumed that the costs of specific exploration efforts will be the same whether or not they are successful. For instance, drilling an exploratory well to a certain depth should cost about the same whether or not it yields a viable resource. Table 7 below defines the range of exploration success rates considered. These exploration success rates represent the probability, on average, that a series of exploration expenditures will result in the identification of a viable resource. These rates also denote the number of exploration wells drilled and associates "hit rate." For example, a rate of 50% denotes identification of the geothermal resource on the second exploration well. By comparison, a rate of 10% denotes that a resource is not identified until the 10th exploration well drilled. The costs associated with exploration activities are provided later in this section and include both drilling and non- drilling[63] activities. The exploration success rate is used to determine the total cost incurred before a viable resource is identified. This analysis does not include a scenario in which exploration capital is exhausted and no geothermal resource is found because LCOE cannot be calculated for such a scenario. Clearly, however, such outcomes are possible and perhaps warrant a probability assignment by project developers and investors.

The expected duration of the confirmation drilling phase could also be considered during the FIT payment-setting process. Excluding the costs associated with this phase from consideration in establishing a FIT would likely yield too low a revenue stream for many investors. As described in Section 5.3.4, the confirmation phase is assumed to include continued deep exploratory drilling, the completion of the resource assessment, and the drilling of initial production wells. This phase is assumed to be financed using 100% equity contributions. Without the benefit of a FIT, it is possible that certain development expenditures might be delayed, and the project timeline thus extended, until the project proponent successfully identifies and negotiates a PPA sufficient to enable permanent financing. Such delay makes the project less attractive to investors, who may compensate for this risk by requiring a higher return on invested capital. This translates into a higher LCOE than for a project that is able to proceed through development more efficiently and expeditiously due to the presence of a FIT that provides investors with power plant revenue certainty, even while the resource assessment is being completed.

Table 7. Range and Definition of Exploration Success Rate

Input Category	← Range of Input Values →			
	Low Cost			High Cost
Exploration Success Rate (%)	50%	33%	25%	10%

Sources: Hance (2005); Augustine et al. (2010)

Table 8. Range of Confirmation Phase Duration

Input Category	← Range of Input Values →		
	Low Cost		High Cost
Confirmation Phase Duration	1.5 years	2 years	3 years

Sources: Augustine (2012)

Table 9 below summarizes the range of assumed geothermal project cost inputs, by development phase. These values are based on the aforementioned literature review and are described in more detail in Appendix A. The range of exploration cost represents differing industry opinion regarding the cost of exploration assuming an average 25% success rate. This analysis employs both the range of exploration cost uncertainty and the range of success rate uncertainty.

Table 9. Range of Geothermal Cost Inputs

Cost Category	←Range of Input Values for Cost Uncertainty[64]→		
	Low Cost		High Cost
Exploration Costs ($/kW)	$100	$175	$225
Confirmation Costs ($/kW)	$300	$550	$750
Site Development Costs ($/kW) (well-field + power plant construction)	$2,000	$3,000	$4,000
TOTAL	$2,400	$3,725	$4,975
Fixed O&M ($/kW-yr)	$40	$50	$60
Variable O&M (cents/kWh)	2.25	3.50	4.50

Sources: Exploration Costs: Sison-Lebrilla and Tiangco (2005), Hance (2005), Salmon et al (2011) Confirmation Costs: Sison-Lebrilla and Tiangco (2005), Hance (2005) Site Development Costs: Sison-Lebrilla and Tiangco (2005), Hance (2005), Geothermex (2010), Augustine et al. (2010)

Fixed O&M: Sison-Lebrilla and Tiangco (2005), Hance (2005), O'Donnell et al. (2009)
Variable O&M: Hance (2005), O'Donnell et al. (2009), Lovekin and Pletka (2009), Augustine et al.(2010), Geothermex (2010).

Table 9 also includes operations and maintenance (O&M) cost estimates. This study's literature review finds that fixed and variable O&M cost categories are often inadequately defined, and that the two categories frequently overlap with each other from study to study. According to O'Donnell et al. (2009), fixed expenses include power plant operations and maintenance while variable O&M includes field, general O&M, and rework; make-up wells; and relocation injection wells. The O&M cost estimates are based on the literature review more broadly, but may be best aligned with the allocation of costs described by O'Donnell. Additional detail on all cost estimates can be found in Appendix A.

5.4.2. Financing Inputs

Beginning in 2008, the global financial downturn shrank the pool of investors available to renewable energy projects. The decline in federal income tax liability associated with deep recession triggered a reduction in the number of viable tax equity investors—those able to effectively monetize the substantial tax benefits generated by renewable energy project investments. Despite this trend, this analysis assumes the continued availability of the Investment Tax Credit[65] (ITC) and assigns equity rates of return that are intended to represent the investor's efficient use of both cash and tax benefits.

At the same time, the economic downturn triggered among lenders the perception of increased market risk. This resulted in a shortening of loan tenors (durations), an increase in lender debt service coverage requirements, and an overall decline in the amount of debt offered to any individual project (Salmon et al. 2011). Debt remains available at the construction stage, with lenders requiring a 25%-45% equity contribution (Salmon et al. 2011). The inputs to this analysis are intended to reflect the current availability of debt, consistent with the literature review. The aforementioned enhancements to the geothermal CREST model account for perceptions of market risks by differentiating capital sources and return requirements by development phase. The table below summarizes the range of financing inputs.

Table 10. Range of Geothermal Financing Inputs, by Development Phase

Project Phase	Capital Structure	← Range of Input Values for Return or Interest Rate →		
		Low Cost		High Cost
Exploration	100% Equity	50%	100% total return	2X (200%)
Confirmation	100% Equity	15%	30% annual return	45%
Site Development (Construction Financing)	35% Equity 65% Debt	15% 5.5%	Equity @ 20% annual return Debt @ 7.5%[2] interest rate	25% 10%
Site Development (Permanent Financing)	45% Equity[1] 55% Debt	10% 5.5%	Equity @ 15% annual return Debt @ 7.5%[2] interest rate	20% 10%
DSCR[66] 1.60 times coverage *Financing Term*: 18 yrs *PPA Term*: 20yrs				

Sources: Salmon et al. 2011; Hance 2005; REFTI 2011

1) Includes cash, tax and developer equity

2) Use of the same interest rate assumption for both construction and permanent debt represents the balancing of two different risk profiles. Construction debt bears the risk of non-completion, but over a short time period. Permanent debt bears the risk of operations, but over a much longer time period. For this analysis, these risks and terms are assumed to balance and market interest rates for both activities are assumed to be the same.

5.4.3. Other CREST Inputs Required for Calculating LCOE

Using CREST for this LCOE analysis requires defining additional assumptions about the plant's specifications and operation. These estimates can be found in Appendix B along with additional detail on the literature review. As with the estimation of installed costs, these assumptions are intended to provide a reference case rather than be representative of any specific operating or proposed facility.

Among these additional inputs is the assumption that the modeled geothermal project will operate for 30 years, but receive payment through a FIT only through the first 20 years. Rather than ignore the revenue potential of the last 10 years, the CREST model allows the user to include an electricity price forecast that it then applies to any gap between the assumed FIT contract duration and the project's useful life.[67] For this analysis, an illustrative electricity market price forecast is provided in order to account for project revenue for years 21 through 30, inclusive. The market price trajectory is calculated using a year-one rate of $40/MWh (in 2012) and an escalation rate of 2.0% (and again, is generic and does not represent actual electricity price projections of any particular location).

5.4.4. Explaining the Range of Possible Inputs

The range of input values presented above is insufficient to provide FIT payment-setting insight on its own. These values provide the foundation for a sensitivity analysis that quantifies the change in LCOE with variation in selected key inputs. The range of financing inputs represents—for illustrative purposes—the potential reduction in cost of capital associated with the establishment of a FIT, which can increase investor certainty and translate into reduced risk. Of course, the magnitude of reduction of any input will be situation specific, personnel specific, and difficult to project. However, the calculation of LCOE over a wide range of plausible inputs, selected from detailed cost studies, reveals which inputs have the greatest impact on LCOE and therefore the greatest risk reduction opportunities. This analysis might help policymakers consider which FIT policy design options are most attractive, depending on their FIT policy goals.

5.5. Reference Case Results

A reference case was derived by modeling LCOE in CREST using the central estimate of each input value described above, and a 25% exploration success rate. The reference case is valuable for comparison purposes only; its LCOE is neither intended to reflect the cost of any specific project nor represent a recommended contract rate. The calculated LCOE for the reference case was 10.85 cents/kWh.

5.6. Sensitivity Analyses: LCOE Results and Lessons Learned

The purpose of this analysis is not to determine the precise cost of any given geothermal system, but rather to demonstrate the impact that a FIT might have on geothermal LCOE.

This is done by changing key inputs from the reference case one or two at a time in order to understand how each factor impacts the cost of geothermal electricity. With these impacts measured, policymakers can gain insight into how a FIT might effectively decrease some of the risks that contribute to the LCOE. Policymakers can also gain insight into how to differentiate FIT payments between areas where geothermal has already been proven and areas where it has not. The former may be assumed to have a higher exploration success rate and therefore a lower LCOE and a lower required FIT payment to attract investment.

We conducted a series of sensitivity analyses[68] that explore changes in LCOE by varying:

- The exploration success rate
- The duration of the confirmation stage
- The FIT contract duration
- Total project cost
- Investor return requirements (equity and debt, if applicable) by development phase.

The result of these sensitivity analyses is a range of geothermal LCOE extending from 8.65 to 13.85 cents/kWh, relative to the reference case LCOE of 10.85 cents/kWh.

The impact on LCOE due to a reduction in soft costs and transaction costs that might result from the availability of a FIT, as discussed in Section 4, could also be examined. While we have not attempted to quantify such costs herein, the impact of reducing these costs, which ultimately reduce the cost subject to permanent financing, is embedded in the examination of the sensitivity of LCOE to total project cost.

5.6.1. Exploration Success Risk

Changes in exploration success rate and return requirements represent the impact on LCOE of a range of investor risk perception with respect to the earliest phase of geothermal development. If the presence of a FIT can provide the earliest investors with a greater degree of confidence that a geothermal resource, if validated, will result in a financed and constructed project, then such investors may be willing to make their capital available with lower expected return requirements to reflect the later reduction in risk due to the presence of a guaranteed long-term contract. If achieved, this reduction in the cost of capital can be translated into lower FIT payments.

The reduction in perceived risk by exploration investors assumes that some projected quantity of exploration expenses are explicitly reflected in the payment-setting process, and that the exploration success rate assumptions used to set the FIT payment align with the geothermal developers own view of a project's development prospects. Consider an example in which two developers, A and B, expect the same cost for exploration activities, but have differing views of expected success rate. Developer A assumes that six exploration wells will need to be drilled in order to identify one productive well. Developer B has a different market outlook and instead assumes that only three wells will be necessary before a "hit" is achieved (a 33% success rate). These two developers have different views on their probability of success, and thus on geothermal LCOE. If the policymaker sets the FIT payment by assuming that one out of every two exploration wells yields a productive resource (a 50% success rate), then neither developer will find it attractive to invest unless the presence of the FIT changes their market outlook. If the FIT payment is established assuming a cost equivalent to four exploration wells for every one productive well (a 25% success rate), then Developer B may find it attractive to invest while Developer A may not. If the FIT payment is established assuming a cost equivalent to seven exploration wells for every one productive well (a 14.3% success rate), then both developers may find it attractive to move forward with project development. This connection between exploration success and FIT payment-setting is illustrated in Figure 4.

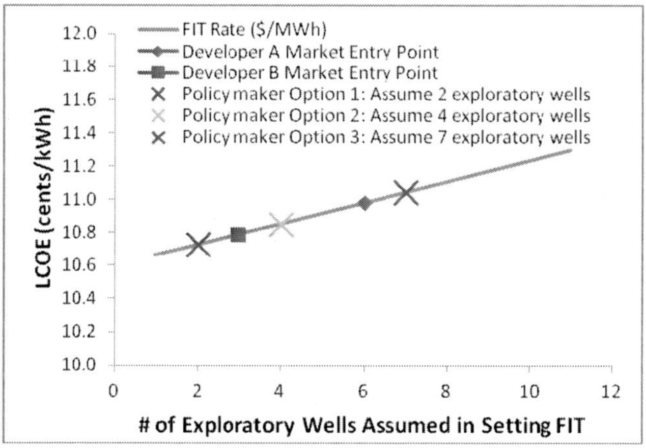

Figure 4. Illustration of connection between exploration success and FIT payment setting.

Figure 4 suggests that the more exploration costs are built into the FIT, the higher the LCOE. This view is incomplete, however, as it ignores the expectation that the inclusion of more exploration cost in the FIT might lower the perceived risk to exploration investors. While difficult to quantify, it is important to consider how increasing the amount of exploration cost in the FIT, combined with guaranteed access to a cost-based revenue stream might reduce the expected return. Using a simplified example, a FIT that assumes it necessary to drill two exploration wells (a 50% success rate) might attract investors with a 200% return requirement. A FIT that assumes it necessary to drill seven exploration wells (a 14% success rate) might attract an exploration investor with a 50% return requirement. By making the act of geothermal exploration far less speculative, it is possible that the required LCOE to attract investment may reflect a lower exploration return. This interaction might cause the line in Figure 4 to flatten out as the number of exploratory wells assumed in the FIT increases, rather than increase linearly as currently shown. In other words, the cost of capital might decrease as the overall probability of success increases.

5.6.2. Confirmation Drilling Risk

Like exploration activities, confirmation drilling, which includes continued deep exploratory drilling, resource assessment, and initial production wells and therefore contains many exploration-type risks, is equity funded in today's market. While it is possible that lenders will show willingness to bear resource risk in the future, they are currently considering investment only in projects that have demonstrated 100% of the projects' required thermal resource. As a result, the cost and availability of confirmation capital is expected to behave much like exploration capital and be determined by the resource and project risk perceived by equity investors and the duration of time investors believe their capital will be exposed to these risks. Where resource risk is perceived to be relatively high and confirmation activities are expected to be spread over three or more years, the cost of capital available to fund such activities increases dramatically. Where resource risk is assumed to be lower and confirmation activities expected to take not much more than a year (as in the expansion of an existing facility), the capital required to fund this aspect of development is expected to be less costly and easier to secure. For this reason, the impact of the duration of the confirmation period and the cost of confirmation capital on LCOE are considered together in this analysis.

Figure 5 shows the impact on LCOE of changes in both confirmation drilling duration and cost of capital together. Confirmation drilling started at a base case of 2 years, and ranged from a low cost sensitivity of 1.5 years, to a high cost sensitivity of 3 years. The cost of capital inputs are detailed in Table

11 and included varying levels of equity rate of returns (as low as 10% for low- cost permanent financing up to 200% for high cost exploration), and debt interest rate (5.5% for low cost and up to 10% for high cost). While the ratio of debt to equity ratio is held constant, it differs by project development stage, based on when debt lenders are interested in financing project development. For example, exploration and confirmation drilling are too risky for debt lenders, so these phases are assumed to be 100% equity financed.

The example shows that a "lower return requirement and shorter contract duration" case (that uses 1.5-year confirmation phase duration and a 15% annual return requirement), results in a required LCOE of about 9.25¢/kWh. When a 3-year confirmation phase duration and 45% annual return requirement are assumed in the "higher return and longer contract duration" case, the required LCOE reaches almost 14¢/kWh. This impact is driven by both the magnitude of the confirmation stage costs (roughly three times greater than exploration costs according to Table 9) and the influence of compounding those costs over time.

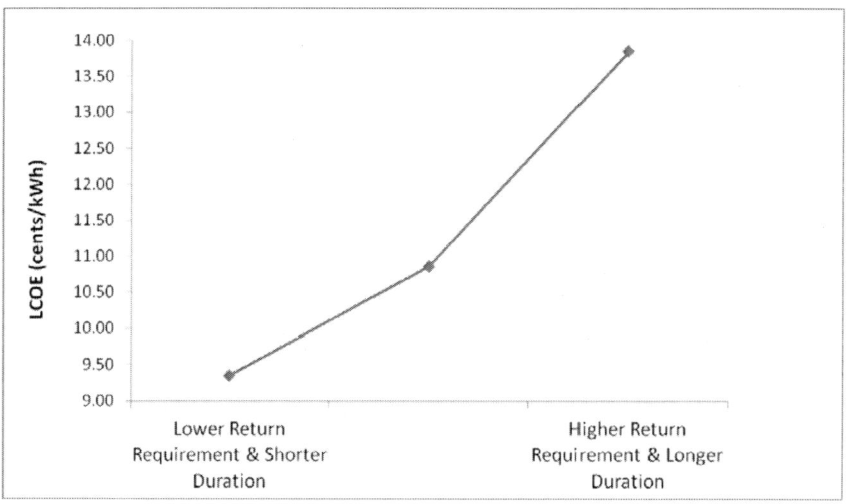

Figure 5. Sensitivity of LCOE to change in both confirmation duration and return requirement.

When investors have confidence in the availability of a financeable contract at a known payment level, it is more likely that they will proceed rapidly with early development activities, shortening the non-technical aspects of the geothermal development cycle, which in turn might reduce LCOE. As a result, if a FIT were able to create investor confidence and shorten the time necessary to attract and deploy capital for development activities, then project LCOE might be reduced.

If by providing the guaranteed availability of a financeable contract at known payment level, a FIT could enhance investor confidence resulting in (1) shortened development times (increased likelihood of proceeding expeditiously with confirmation after exploration yields potentially viable resource) and (2) lowered perceived risk (reflected in lower threshold equity cost of capital expectations), then a FIT could have the potential to substantially reduce the required LCOE to attract investment.

5.6.3. FIT Contract Duration

Longer contract durations facilitate lower LCOEs by extending the period over which equity investments and debt obligations are repaid. Through FITs, this benefit is realized as long as the contract price is known for the duration of the agreement. This report and quantitative analysis assumes a 20-year FIT throughout. The sensitivity of LCOE to FIT duration was also tested for FITs of 10, 15, 25, and 30 years. Geothermal projects operating beyond the FIT term are assumed to collect (forecasted) wholesale market revenues. No other payments are assumed during this period.

The impact of FIT contract duration on LCOE is shown in Figure 6. As shown, LCOE is highest for shorter duration FIT contracts, and the total impact between 10 and 30 years is approximately 15%. The impact on LCOE levels out as the FIT contract duration extends and it virtually flattens out for FIT payments of 25 and 30 years. These results may illustrate why most FIT policies are between 15–20 years—most of the benefit is realized in this time period and extending the FIT longer than 20 years does not provide a marked additional benefit in reducing the LCOE. However, a longer FIT duration may expand the pool of investors interested in geothermal project development, a factor that is difficult to capture quantitatively.

5.6.4. Installed Cost and Investor Return Requirements

The cost of geothermal electricity is influenced by the installed project costs, as well as by the availability and cost of permanent financing. Installed project costs were detailed in Table 9, and ranged from \$2,400/kW for the low cost case to \$4,975/kW in the high case. These cost estimates are assumed to include all permanent financing transaction costs (e.g., bank fees and reserve accounts) as well as payments to "cash out" the investors who supported all previous development phases. To this end, total project costs are impacted by the return requirements of each development stage. If the surety of a FIT reduces the cost of capital at any point in the development process, the impact of this should be felt, albeit potentially modestly, as a reduction in total project cost.

Permanent financing assumptions are shown in Table 11. This is comprised of the long-term debt and equity that enable the payoff of construction financing and fund the ongoing operation of the geothermal project; their required return is expected over an extended period of time. For debt, this period is most likely to match (or be a few years shorter than) the length of the FIT. For equity, the investment period is the project's useful life, unless a sale of the asset is negotiated at an earlier date. To illustrate the impact of cost of capital on LCOE for a range of different project installed costs, the costs of equity and debt are combined into the weighted average cost of capital (WACC), a proxy for the total cost of permanent financing.

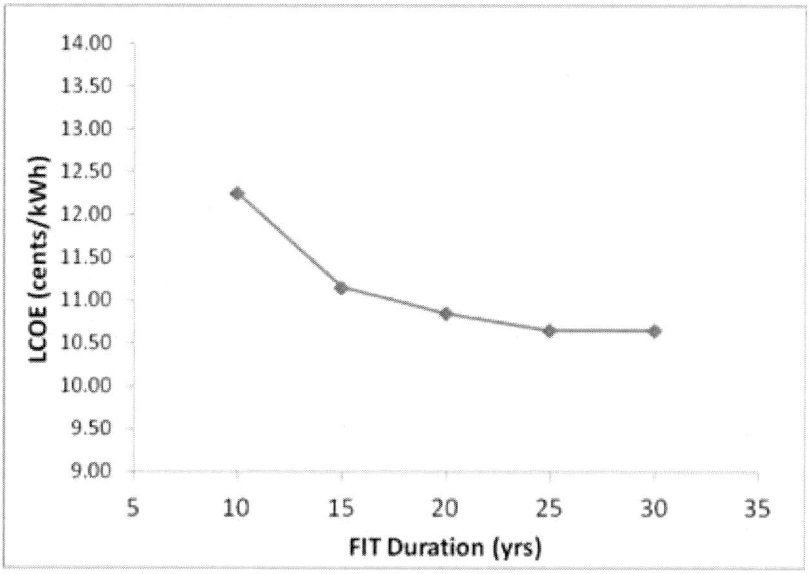

Figure 6. Illustration of relationship between contract duration and LCOE.

Table 11. Cost of Capital Assumptions for Lower and Higher Cost Sensitivity Analyses

	Lower Cost of Capital		Higher Cost of Capital
Permanent Equity (includes cash, tax, and developer equity)	10%	15%	20%
Permanent Debt (18-year tenor on 20-year PPA)	5.5%	7.5%	10%
Weighted Average Cost of Capital (50/50)	**6.6%**	**9.7%**	**13.0%**

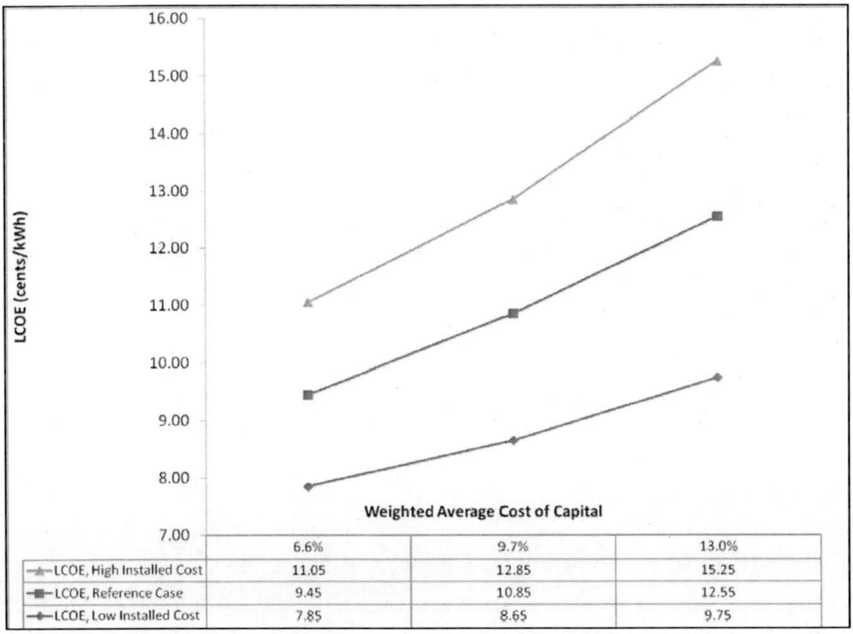

Figure 7. Sensitivity of LCOE to change in cost of permanent financing, shown at three different levels of installed cost.

Figure 7 shows the impact of the WACC as well as the capital cost on LCOE. As shown, the range of possible installed costs (Low = $2,400/kW, Reference = $3,725/kW, and High = $4,975/kW) greatly impacts the LCOE. Figure 7 illustrates three important points:

- For any given installed cost, the WACC impacts the LCOE significantly; if a FIT can reduce perceived risk, this relationship suggests that a FIT might materially reduce LCOE. For example, if the presence of a FIT enabled a 100 basis point savings on the WACC, the Reference Case in this analysis would be reduced from 10.85 to 10.45¢/kWh (a reduction of 3.7%).
- If a FIT could reduce or eliminate various cost components, such as contract price risk premiums, bid security, legal, bid preparation, other transaction costs, or the cost of capital at either the exploration or confirmation phases, the LCOE would drop accordingly. For instance, a 5% drop in installed cost for a reference-level installed cost project would yield a roughly 0.3¢/kWh reduction in LCOE.
- The higher the installed capital cost, the greater the impact on LCOE of reducing the WACC.

5.7. Summary of LCOE Sensitivity Results

Figure 8 summarizes each sensitivity analysis by showing the range of FIT payment outcomes associated with the inputs tested (x-axis), and ranking their impacts on FIT payment relative to one another (y-axis). The input with the greatest impact on FIT payment occupies the highest position on the diagram. The array of input assumptions associated with these LCOE outcomes are based on the literature review and provided in the tables throughout this section. Additional detail is available in Appendix A.

Importantly, this analysis only considers the potential impact of a FIT policy and does not compare FITs to other policies, and therefore cannot conclude if a FIT is the best policy to achieve all of the desired objectives. Further analysis is needed to compare FITs to other policies for their effectiveness and cost efficiency.

Policies that spur decreases in the cost of capital at any development stage (by increasing investor confidence, increasing the pool of available capital, or other means) and help to reduce development delays are expected to have a sizable impact on LCOE. These impacts are demonstrated particularly at the confirmation phase, as it has been defined in this report. The duration of price-certain, creditworthy contracts (such as FITs) are also expected to have a notable impact on LCOE. This analysis also concludes that there are factors, such as total installed cost, which have the potential to impact LCOE similarly but are only modestly influenced by the presence of a FIT through reductions in fees and transaction costs typically associated with contract identification and negotiation, competitive bidding procedures, and/or permanent financing. Beyond the general nature of FIT policies to increase investor certainty, specific policy design choices regarding the manner in which exploration and confirmation success and return requirements are included in FIT payment-setting are also shown to have a direct and meaningful impact on geothermal LCOE. Each of these outcomes can be achieved, either directly or indirectly, through the establishment of a long-term, price-certain FIT. Policymakers can use this approach to prioritize FIT design characteristics that best support their objectives.

It is important to emphasize that this analysis comprises a "what if" analysis. While the directionality of impacts—the relationship between risk and return—are well-understood, the literature provides limited data on the quantitative impacts of these types of risk reduction on required investor returns. In addition, the specific impacts may differ by investor and at different times and circumstances. While actual FIT impacts on specific project LCOEs are difficult to determine, this analysis demonstrates that reductions in LCOE could result from careful choices in FIT design and implementation.

FITs have the potential to reduce the cost of capital by increasing investor confidence. In addition, the knowledge that a long-term price-certain contract is available for all qualifying generators may also help to avoid delays in the development process and interruptions in funding caused by the search for, and negotiation of, a financeable power purchase arrangement. In this vein, a well-developed FIT policy could result in compounding LCOE reductions. To this end, it is important to consider the role that an established and predictable program for long-term contracts plays in a project's ability to secure the most competitive financing.

While not demonstrable through a quantitative analysis, it is expected that the reduction in perceived risk that generates reductions in LCOE would also increase the pool of investors interested in funding the geothermal sector. The potential result is both equity and debt, which are more accessible to geothermal developers and perhaps on more favorable terms. The results demonstrate the positive impact of lower return requirements and shorter development cycles on geothermal LCOE. Further, the presence of a FIT may drive changes in investment terms for geothermal that perpetuate the potential LCOE differences shown in this analysis. In other words, once FITs are shown to reduce development risk, contracting risk, and the perception of overall market risk, it is increasingly likely that reductions in investor return requirements will follow as competition for geothermal investments increases, continuing the trend shown in this sensitivity analysis.

6. CONCLUSIONS

Geothermal electricity represents a significant and underutilized source of renewable electricity in the United States. Geothermal power plant developers face challenges beyond those experienced by other renewable electric facility developers. This analysis focuses on ways to address risks other than the physical risk of drilling unproductive exploratory wells, particularly through the design and implementation of geothermal-specific FIT policies. The unique characteristics of geothermal electricity development have important implications for the design of effective policy frameworks and incentives intended to support geothermal exploration and production.

6.1. International Experience Provides Opportunity to Tailor Geothermal FITs to U.S. Market

Despite international interest in FITs, there has been a limited focus on geothermal-specific FIT design. Sixteen countries have established specific FIT incentives for geothermal, but most are relatively new or limited in scope (e.g., project size limitations). FITs have been introduced that not only target geothermal development, but that target small scale plants, combined heat-and- power applications, specific types of geothermal resources, and specific geographies. While performance-based incentive structures have played an important role in the growing U.S. renewable energy industry, they have not provided the long-term certainty and basis for financing necessary to create a new wave of geothermal development. By studying international tariffs and policy design choices, this report concludes that geothermal FITs can help facilitate the development of geothermal power plants by the private sector, when they are carefully designed and especially for proven resources. Because geothermal FITs are relatively new, more analysis could help target FITs to the unique risks of geothermal power projects.

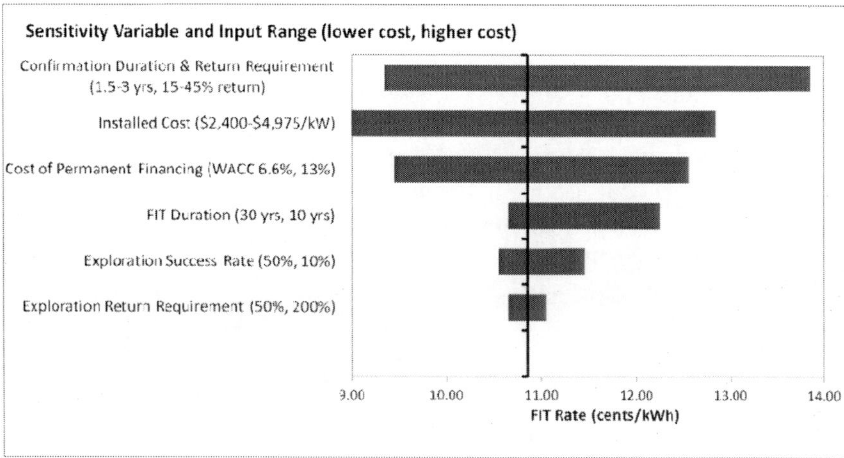

Figure 8. Relative impact on LCOE (and thus FIT Payment Level) of each sensitivity analysis.

6.2. FIT Policy Design Provides Opportunities to Support Geothermal Deployment and Attract Additional Investment Capital into Geothermal Exploration and Development Activities

The report finds that the key issues relating to geothermal FITs in the United States include:

- Linking FITs to specific geothermal targets to demonstrate a state's commitment to geothermal procurement
- Targeting the FITs to specific types of geothermal resources (e.g., hydrothermal versus EGS) and differentiating payment levels by technology to meet investors' required returns
- Incorporating exploration expense and risk in FIT payment-setting methodology and payment levels to attempt to increase investment in inherently risky activity of geothermal exploration
- Creating standard pricing and standardized off-take contracts to reduce risk and transaction costs
- Structuring and setting the FIT payment to reduce risk
- Reducing market price risk by targeting appropriate contract duration
- Considering longer-term, stable FIT policies and FIT contracts

- Indexing payments to inflation, drilling costs or other relevant indicators
- Exploring scheduled and transparent policy adjustments over time.

We conclude that the availability of a FIT is likely to increase the likelihood of successful geothermal power plant development, financing, and operation relative to situations in which long-term, price-certain contracts are not guaranteed to qualifying generators. Importantly, the analysis did not complete a comprehensive analysis of ways to mitigate exploration risks using FITs or other policies.

6.3. FIT Policies can be Designed to Address Key Issues that Have a Significant Impact on LCOE: Confirmation Phase Duration and Equity Cost, Cost of Permanent Financing, and FIT Duration

The assurance of a long-term, price-certain contract is expected to provide geothermal projects with access to a broader spectrum of less-expensive capital. If the cost of capital can be reduced at each stage of development, reductions in LCOE will result. When establishing return requirements, investors will consider both the nature of the risk and the expected duration that their capital will be outstanding. Longer development and/or contracting time horizons translate into greater risk and a higher target return requirement for the pre-operating phase. By comparison, as the FIT contract duration increases, the LCOE is expected to decrease as debt and equity investments can be recovered over a longer period of time while maintaining revenue certainty. The illustrative quantitative analysis in Section 5 identifies the duration of the confirmation phase, the cost of confirmation phase equity, and the cost of permanent financing and FIT duration as the highest-impact inputs that can be influenced by FIT policy design.

The availability of a long-term, price-certain FIT contract is guaranteed upon commercial operation and can help to shorten the geothermal development cycle by removing a lengthy and uncertain negotiation for a competitive PPA. In this case, significant benefits can be realized in the form of a lower LCOE. Similarly, if the availability of a FIT can increase the pool of available investment capital, then both competitive forces and a real reduction in the risk profile of geothermal investments is expected to result in a lower cost of capital for permanent equity and debt. This analysis demonstrates the impact of such a trend in the form of lower LCOEs. Ultimately, lower LCOE would mean a lower FIT payment level, lower geothermal project development costs, and lower overall costs to ratepayers.

APPENDIX A. SUMMARY OF LITERATURE REVIEW ON U.S. GEOTHERMAL COSTS

The cost, performance, and financing inputs employed in the LCOE analysis were developed through a detailed literature review. Publicly available geothermal cost research was compared and used to develop the low case, reference case, and high case estimates for the LCOE modeling. Cost comparisons and case definitions are shown in Figures A-1 through A-6.

(All data adjusted to 2010$; shown on graph in year of study)

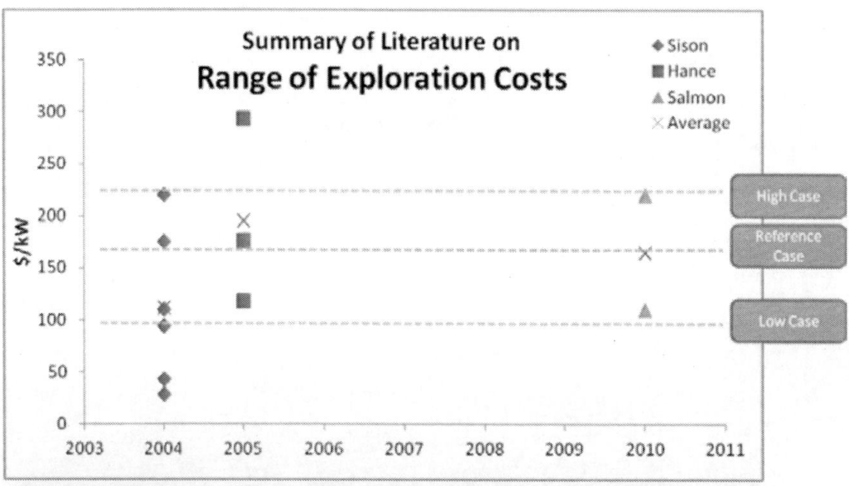

	Reference Case	Low Sensitivity	High Sensitivity
Exploration Costs Modeled ($/kW)	$175	$100	$225

Figure A-1. Range of exploration costs.

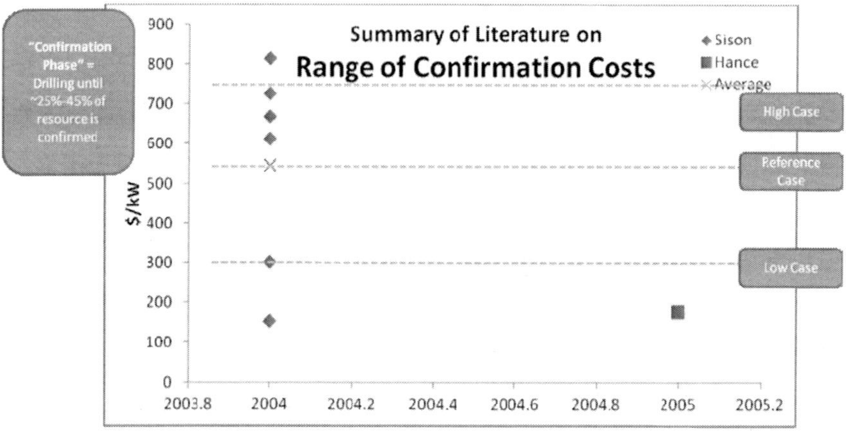

	Reference Case	Low Sensitivity	High Sensitivity
Confirmation Costs Modeled ($/kW)	**$550**	$300	$750

Figure A-2. Range of confirmation costs.

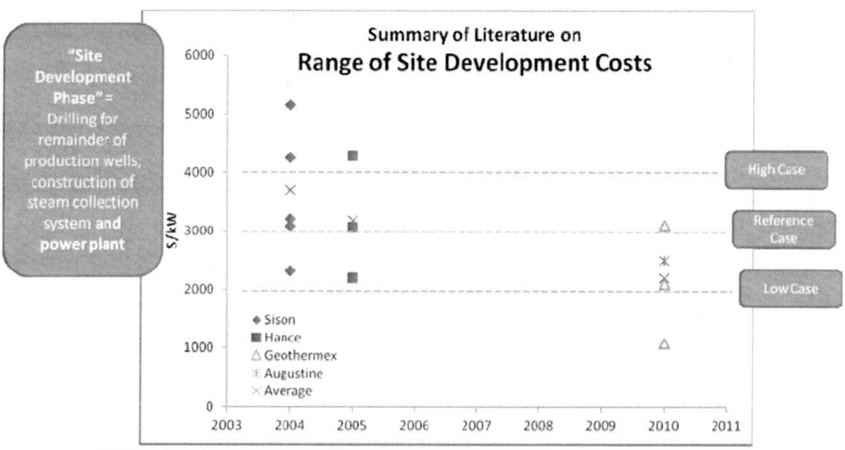

	Reference Case	Low Sensitivity	High Sensitivity
Site Development Costs Modeled ($/kW)	**$3,000**	$2,000	$4,000

Figure A-3. Range of site development and construction costs.

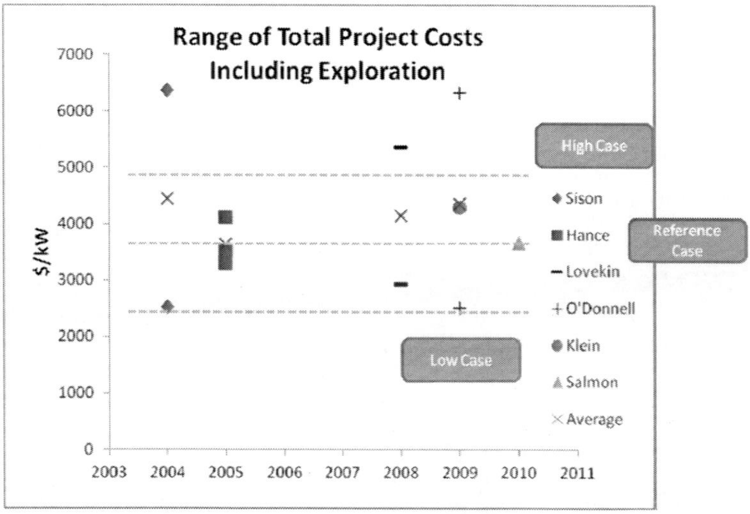

	Reference Case	Low Sensitivity	High Sensitivity
Total Project Costs ($/kW)	**$3,725**	$2,400	$4,975

Figure A-4. Range of total project costs (including exploration).

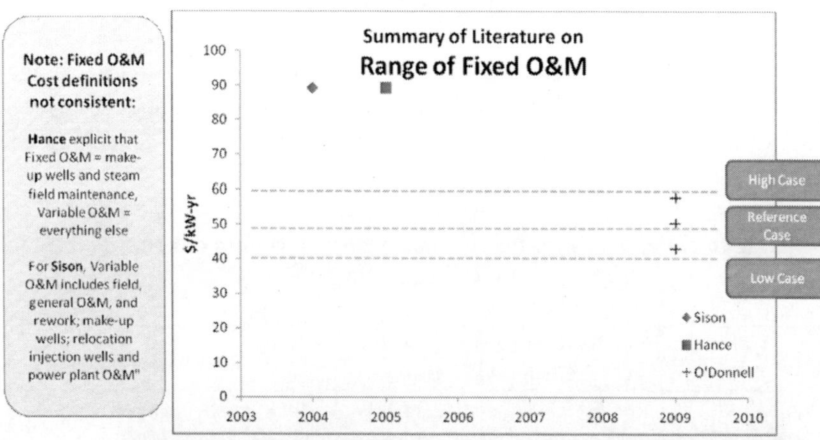

	Reference Case	Low Sensitivity	High Sensitivity
Fixed O&M Costs Modeled ($/kW)	**$50**	$40	$60

Figure A-5. Range of fixed O&M costs.

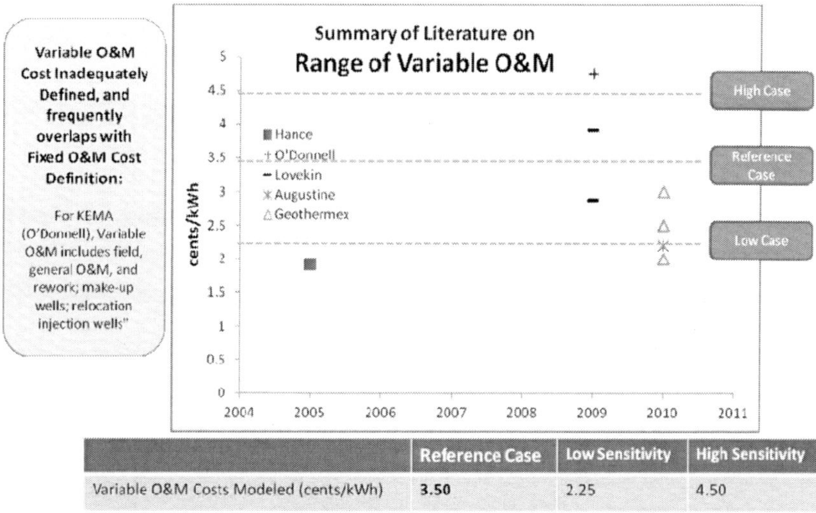

Figure A-6. Range of variable O&M costs.

APPENDIX B. SUMMARY OF ADDITIONAL CREST MODELING INPUTS

Calculating LCOEs requires defining additional assumptions about the plant's specifications and operation. These assumptions are intended to be central estimates rather than representative of any specific operating or proposed facility.

Table B-1. Project Size, Performance, and Development Inputs for LCOE Modeling

Input Category	Reference Case	Potential Range	Notes and Sources
Nameplate Capacity	**50 MW**	2–100 MW	Most values between 15 and 50 MW Hance (2005), O'Donnell et al. (2009), Salmon et al. (2011), Klein (2009)
Net: Gross Capacity Ratio	**80%**		20% for "auxiliary loads"

Input Category	Reference Case	Potential Range	Notes and Sources
Availability Factor	**90%**	80%–95%	Hance (2005), Lovekin and Pletka (2009), O'Donnell et al. (2009), Sison-Lebrilla and Tiangco (2005), Klein (2009)
Project Useful Life	**30 Years**		Hance (2005)
Annual Degradation of Thermal Resource	Assumes that combination of plant sizing and make-up wells keeps plant operating at capacity. Cost captured in O&M estimates.		
Exploration Well Success Rate	**25%**	10%-50%	20%-25% Hance (2005) 35% for "Hydro" Augustine et al. (2010)
Exploration Period Duration	**3 Years**	2–4 Years	NREL professional analysis experience
Confirmation Well Success Rate	**60%**		Hance (2005)
Confirmation Period Duration	**2 Years**	1.5–3 Years	NREL professional analysis experience
Production Well Success Rate	**75%**		NREL professional analysis experience
Construction Duration (Well Field and Power Plant)	**2 Years**		NREL professional analysis experience
Royalties	**4%**		Hance (2005)

APPENDIX C. GEOTHERMAL POWER UNDER PURPA

The Public Utilities Regulatory Policy Act is a federal law passed as part the broader National Energy Act (NEA) of 1978. Against the backdrop of the oil crises of the 1970s, during which energy prices tripled in 1974, the goal of NEA was to decrease national dependence on imported oil. The goal of PURPA was to increase the amount of electricity purchased from independently- owned renewable energy and cogeneration plants. PURPA required utilities to a) purchase energy and/or capacity from non-utility generators[69] and b) purchase the commodities at rates based on utilities' avoided cost. In order to be eligible under PURPA, renewable energy generators had to be less than 50% owned by utilities, smaller than 80 MW of installed capacity, and have at least 75% of the total energy input provided by renewable energy (e.g., in the case of biomass plants co-fired with fossil fuels). PURPA rapidly expanded non-utility generation in the United States. By 1996, PURPA had supported the development of approximately 12,600 MW of renewable power, of which 11% was geothermal capacity in three states (Guey-Lee 1999). The majority of the geothermal capacity supported by PURPA was concentrated in California, as can be seen in Figure 15 below.

The Federal Energy Regulatory Commission, which was responsible for certifying eligible generators under PURPA, left the definition of utility avoided cost to each state. Different states defined avoided cost in different ways. California created several standard offer power purchase contracts that generators could choose from. Standard Offer No. 4 (SO4), which defined avoided cost as a 10-year schedule of escalating payments, drove a boom in renewable energy development (Pierce and Livesay 1994). California supported approximately 25% of all renewable energy generation under PURPA, of which the largest percentage (~40%) was geothermal (Guey-Lee 1999). The price paid under PURPA for geothermal varied widely from state to state. The average price paid for geothermal plants in California was 12.44 ¢/kWh, for example. By contrast, the average price paid in Nevada was 5.31 ¢/kWh (Guey-Lee 1999).

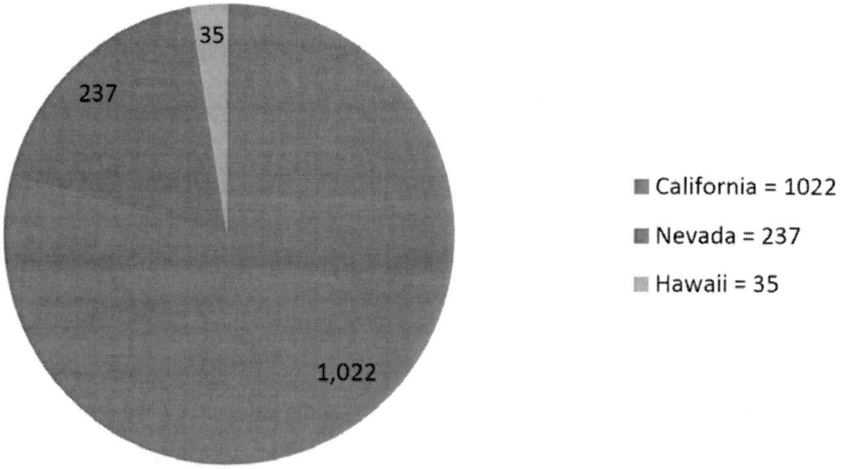

Source: Guey-Lee (1999).

Figure C-1. Installed geothermal capacity under PURPA (MW).

Nationwide, PURPA stimulated exploration drilling at more than 50 prospects between 1979-1985, many of which were subsequently developed. Many of the geothermal sites developed under PURPA benefitted from federal cost-share programs, including both grants for exploration and confirmation drilling and loan guarantee programs (DOE 2008). The remaining exploration activity was primarily financed through equity from large oil, gas, mining, and utility companies (GeothermEx 2010). As the U.S. DOE's Geothermal Technologies Program Blue Ribbon Panel concluded, "Private industry was willing to take more drilling risk because of the higher electricity price" (EERE 2011, p. 8), in addition to the other programs and incentives.

Following the initial renewable energy market growth under PURPA, a decline in the price of oil, the entry of lower-cost natural gas power generation into the electricity market, concerns over excess generating capacity, and a sharp drop in the cost of commercial loans dramatically decreased the avoided cost of most utilities. This not only reduced the prices offered under PURPA but also created controversy about the pricing of existing PURPA contracts. By 1995, the average wholesale price of electricity was 3.53 ¢/kWh, whereas the average price for non- utility generators under PURPA was 6.31 ¢/kWh (Guey-Lee 1999). PURPA was amended in 2005 to remove the mandatory purchase obligation and requirement that utilities pay avoided cost-based rates in states where generators have "nondiscriminatory access to competitive markets or open-access transmission services provided by a regional transmission operator" (Elefant 2011). While utilities in some jurisdictions have applied for these PURPA exemptions, utilities in other states have continued to procure renewable energy under PURPA. In Idaho, for example, PURPA supported the development of the Raft River geothermal plant, which came online in 2008 (Doris et al. 2009).

PURPA was similar to modern FITs in that it allowed IPPs to connect and sell power to the grid. PURPA has also been crediting with not only jumpstarting renewable energy development, but also inspiring FITs in countries such as Denmark and Germany. Unlike many modern FITs, however, the price paid under PURPA was based on avoided cost, rather than on the generation cost of specific technologies.

References

Augustine, C. (28 November and 5 December 2011). Personal communication. National Renewable Energy Laboratory, Golden, CO.

Augustine, C., Young, K. R. & Anderson, A. (February 2010). *Updated U.S. Geothermal Supply Curve.* NREL/ CP-6A2-47458. Golden, CO: National Renewable Energy Laboratory. Accessed May 1, 2012: http://www.nrel.gov/docs/fy10osti/47458.pdf.

Barbose, G. (23 February 2011). *State RPS Policies in the United States.* Proceedings of the Presentation to the Korea Delegation.

Beerepoot, M. (2011). "Technology Roadmap: Geothermal Heat and Power." Paris, France: International Energy Agency. Accessed May 22, 2012: www.iea.org/papers/2011/Geothermal_Roadmap.pdf.

Bertani, R. (May 2005). "World Geothermal Power Generation 2001-2005." *Geothermics, the International Journal of Geothermal Research and its Applications*, (*34*:6), 651-690. Accessed May 1, 2012: http://geothermal.org/articles/worldpower05.pdf.

Bertani, R. (25-29 April 2010). *Geothermal Power Generation in the World: 2005-2010 Update Report.* Proceedings of the World Geothermal Congress 2010, Bali, Indonesia. Accessed May 1, 2012: http://geothermie.nl/fileadmin/user_upload/documents/bestanden/IGA/Geothermal_electricity_in_the_world_2010_report_Ruggero_Bertani.pdf.

Boonin, D. M. (December 2008). *Feed-In Tariffs: Best Design Focusing Hawaii's Investigation.* Washington, DC: National Regulatory Research Institute. Accessed May 1, 2012: http://www.wind-works.org/FeedLaws/USA/Hawaii%20NRRC%20Feed-In%20Tariffs%20from%20PUC-121108.pdf.

BMU. (June 2003). *Aktualisierung des Berichtes über den Stand der Markteinführung und der Kostenentwicklung von Anlagen zur Erzeugung von Strom aus erneuerbaren Energien (Erfahrungsbericht zum EEG) vom 28. Juni 2002 durch das Bundesministerium für Umwelt, Naturschutz und Reaktorsicherheit.* Berlin, Germany: Bundesministerium für Umwelt Naturschutz und Reaktorsicherheit. Accessed May 1, 2012: http://www.loy-energie.de/download/A-Drs%2015-15-121.pdf.

BMU. (November 2007). *Erfahrungsbericht 2007 zum Erneuerbaren-Energien-Gesetz (EEG) gemäß § 20 EEG.* Berlin, Germany: Bundesministerium für Umwelt Naturschutz und Reaktorsicherheit. Accessed May 1, 2012: http://www.bmu.de/files/pdfs/allgemein/application/pdf/erfahrungsbericht_eeg_2007.pdf.

BMU. (2011a). *Erfahrungsbericht 2011 zum Erneuerbare-Energien-Gesetz (EEG- Erfahrungsbericht).* Berlin, Germany: Bundesministerium für Umwelt Naturschutz und Reaktorsicherheit. Accessed May 1, 2012:

http://www.bmu.de/files/pdfs/allgemein/application/pdf/eeg_erfahrungsbericht_2011_bf.pdf.

BMU. (2011b). RES LEGAL. Berlin, Germany: Bundesministerium für Umwelt Naturschutz und Reaktorsicherheit. Accessed November 11, 2011: http://www.res-legal.de/en.html.

Caperton, R. W., Hendricks, B., Lauer, J. & Hight, C. (2011). *CLEAN Contracts: Making Clean Local Energy Accessible Now*. Washington, DC: Center for American Progress, Energy Action Coalition, and Groundswell.

Ciesielska, J., Concas, G., Despotou, E., Fontaine, B., Garbe, K. & Montoro, D. F. (2011). *Global Market Outlook for Photovoltaics Until 2015*. Belgium: European Photovoltaic Industries Association. Accessed May 1, 2012: http://www.epia.org/publications/epiapublications/global- market-outlook-for-photovoltaics-until-2015.html.

Corfee, K., Karcher, M., Cleijne, H., Burgers, J., Faasen, C., Tong, N., Rickerson, W., Grace, R. & Gifford, J. (August 2010). *Feed-In Tariff Designs for California: Implications for Project Finance, Competitive Renewable Energy Zones, and Data Requirements*. Sacramento, CA: California Energy Commission. Accessed May 1, 2012: http://www.energy.ca.gov/2010publications/CEC-300-2010-006/CEC-300-2010-006.PDF.

Cory, K. (6 June 2011). "FERC Ruling Changes Course as Assists Renewables." National Renewable Energy Laboratory, Renewable Energy Project Finance. Accessed, August 18, 2011: http://financere.nrel.gov/finance/content/ferc-ruling-changes-course-and-assists-renewables.

Cory, K., Couture, T. & Kreycik, C. (March 2009). *Feed-In Tariff Policy: Design, Implementation and RPS Policy Interactions*. NREL/TP-6A2-45549. Golden, CO: National Renewable Energy Laboratory. Accessed May 1, 2012: http://www.nrel.gov/docs/fy09osti/45549.pdf.

Couture, T. (2011). "Booms, Busts, and Retroactive Cuts: Spain's RE Odyssey." *E3 Analytical Brief* (3:1); pp. 1-4. Accessed May 22, 2012: http://e3analytics.ca/documents/Analytical_Brief_Vol3_Issue1.pdf.

Couture, T. & Cory, K. (June 2009). *State Clean Energy Policies Analysis (SCEPA) Project: An Analysis of Renewable Energy Feed-In Tariffs in the United States*. NREL/TP-6A2-45551. Golden, CO: National Renewable Energy Laboratory. Accessed May 1, 2012: http://www.nrel.gov/docs/fy09osti/45551.pdf.

Couture, T., Cory, K., Kreycik, C. & Williams, E. (July 2010). *A Policymaker's Guide to Feed-In Tariff Policy Design*. NREL/TP-6A2-

44849. Golden, CO: National Renewable Energy Laboratory. Accessed May 1, 2012: http://www.nrel.gov/docs/fy10osti/44849.pdf.

Couture, T. & Gagnon, Y. (2010). "An Analysis of Feed-In Tariff Remuneration Models: Implications for Renewable Energy Investment." *Energy Policy*, (*38*, 2), 955-965. Accessed May 1, 2012: http://www.e3analytics.ca/documents/fitpolicy.pdf.

Cross, J. & Freeman, J. (2009). *2008 Geothermal Technologies Market Report.* DOE/GO 1020092864. Washington, DC: U.S. Department of Energy, Office of Energy Efficiency and Renewable Energy. Accessed May 1, 2012: http://www.nrel.gov/analysis/pdfs/46022.pdf.

Deutsche Bank Climate Change Advisors. (December 2009). *Paying for Renewable Energy: TLC at the Right Price - Achieving Scale Through Efficient Policy Design.* New York, NY: The Deutsche Bank Group. Accessed May 1, 2012: https://www.dbadvisors.com/content/_media/1196_Paying_for_Renewable_Energy_TLC_at_the_Right_Price.pdf.

Deutsche Bank Climate Change Advisors. (May 2011). *The German Feed-In Tariff for PV: Managing Volume Success with Price Response.* New York, NY: Deutsche Bank Group. Accessed May 1, 2012: https://www.Db advisors.com/content/_media/DBCCA_German_FIT_for_PV_0511.pdf.

de Jager, D. & Rathmann, M. (October 2008). *Policy Instrument Design to Reduce Financing Costs in Renewable Energy Technology Projects.* Utrecht, Netherlands: Ecofys International BV. Prepared for the International Energy Agency, Renewable Energy Technology Development. Accessed May 1, 2012: http://www.ecofys.com/files/files/report_policy_instrument_design_to_reduce_financing_costs_i n_renewable_energy_technology_pro.pdf.

Deloitte. (February 2008). *Geothermal Risk Mitigation Strategies Report.* Washington, DC: U.S. Department of Energy, Office of Energy Efficiency and Renewable Energy, Geothermal Program. Accessed May 1, 2012: http://www1.eere.energy.gov/geothermal/pdfs/geothermal_risk_mitigation .pdf.

U.S. Department of Energy (DOE). (2012). "How Enhance Geothermal Systems Work." U.S. Department of Energy, Geothermal Technologies Program. Accessed August 3, 2012: http://www1.eere.energy.gov/geothermal/egs_animation.html.

DOE. (2008). "Geothermal Technologies Program Multi-Year Research, Development, and Demonstration Plan: 2009-2015 with Program Activities to 2025 (Draft)." Washington, DC: U.S. Department of Energy.

Accessed May 26, 2012: http://www1.eere.energy.gov/geothermal/pdfs/gtp_myrdd_2009-complete.pdf.

Doris, E., Kreycik, C. & Young, K. (September 2009). *Policy Overview and Options for Maximizing the Role of Policy in Geothermal Electricity Development.* NREL/TP-6A2-46653. Golden, CO: National Renewable Energy Laboratory. Accessed May 1, 2012: http://www.nrel.gov/docs/fy10osti/46653.pdf.

EERE. (June 2011). *Geothermal Technologies Program Blue Ribbon Panel recommendations* (Draft). Washington, DC: US Department of Energy, Office of Energy Efficiency and Renewable Energy. Accessed May 1, 2012: http://www1.eere.energy.gov/geothermal/pdfs/brp_draft_report_june_17_2011.pdf.

Electricity Regulatory Authority. (2010). *Uganda Renewable Energy Feed-In Tariff (REFIT) Phase 2: Approved Guidelines for 2011-2012.* Accessed May 1, 2012: http://www.era.or.ug/Pdf/Approved_Uganda%20REFIT%20Guidelines%20V4%20(2).pdf.

Elefant, C. (2011). *Reviving PURPA's Purpose: The Limits of Existing State Avoided Cost Rate Making Methodologies in Supporting Alternative Energy Development and a Proposed Path for Reform.* Washington, DC: The Law Offices of Carolyn Elefant. Accessed May 1, 2012: http://www.recycled-energy.com/images/uploads/Reviving-PURPA.pdf.

EGEC. (2011). *Feed-In Tariffs and Green Certificates for Geothermal in the EU-27.* Brussels, Belgium: European Geothermal Energy Council.

Farrell, J. (October 2011). *CLEAN v SRECs: Finding the More Cost-Effective Solar Policy.* Minneapolis, MN: Institute for Local Self-Reliance. Accessed May 1, 2012: http://www.ilsr.org/author/johnf/page/7/.

Fink, S., Porter, K. & Rogers, J. (October 2010). *The Relevance of Generation Interconnection Procedures to Feed-In Tariffs in the United States.* NREL/SR-6A20-48987. Golden, CO: National Renewable Energy Laboratory. Accessed May 1, 2012: http://www.nrel.gov/docs/fy11osti/48987.pdf.

Fish, J. R. & Heaps, C. M. (27 March 2009). *Fundamentals of Geothermal Financing.* Proceedings of the West Coast Geothermal Finance and Development Workshop, San Francisco, CA.

GEA. (2012). *Annual U.S. Geothermal Power Production and Development Report.* Washington, DC: Geothermal Energy Association. Accessed May 15, 2012: http://geo- energy.org/.

GEOFAR. (2009a). *Financial Instruments as Support for the Exploitation of Geothermal Energy.* Brussels, Belgium: European Commission,

Intelligent Energy Europe, Geothermal Finance and Awareness in European Regions. Accessed May 1, 2012: http://www.energia.gr/geofar/articlefiles/geofar_report_eng2.pdf.

GEOFAR. (2009b). *Non-Technical Barriers and the Respective Situation of the Geothermal Energy Sector in Selected Countries*. Brussels, Belgium: European Commission, Intelligent Energy Europe, Geothermal Finance and Awareness in European Regions. Accessed May 1, 2012: http://www.energia.gr/geofar/articlefiles/2nd_report.pdf.

GEOFAR. (April 2010). *Case Studies for Selected Geothermal Operations*. Brussels, Belgium: European Commission, Intelligent Energy Europe, Geothermal Finance and Awareness in European Regions. Accessed May 1, 2012: http://www.energia.gr/geofar/page.asp?p_id=137&cat_id=19&lng=1.

GEOFAR. (2011). *Emerging Financing Scheme for Fostering Investment in the Geothermal Energy Sector*. Brussel, Belgium: European Commission, Intelligent Energy Europe, Geothermal Finance and Awareness in European Regions. Accessed May 1, 2012: http://www.energia.gr/geofar/articlefiles/geofar_report_eng_web.pdf.

GeothermEx. (June 2010). *An Assessment of Geothermal Resource Risks in Indonesia*. Washington, DC: The World Bank, Public-Private Infrastructure Advisory Facility. Accessed May 1, 2012: http://www.ppiaf.org/sites/ppiaf.org/files/publication/REPORT_Risk_Mitigation_Options_Indon esia.pdf.

German DENA. (2012). "Geothermal Energy - Petrochemical Energy." Deutsche Energie- Agentur (German Energy Agency). Accessed August 3, 2012: http://www.renewables-made-in-germany.com/index.php?id=572.

Gifford, J. S., Grace, R. C. & Rickerson, W. H. (March 2011). *Renewable Energy Cost Modeling: A Toolkit for Establishing Cost-Based Incentives in the United States*. NREL/TP-49143. Golden, CO: National Renewable Energy Laboratory. Accessed May 1, 2012: http://www.nrel.gov/docs/fy11osti/51093.pdf.

Gipe, P. (5 October 2011). "Tables of Renewable Tariffs or Feed-In Tariffs Worldwide." Accessed November 11, 2011: http://www.wind- works.org/FeedLaws/TableofRenewableTariffsorFeed-InTariffsWorldwide.html.

Global Wind Energy Council. (April 2011). *Annual Market Update 2010*. Accessed May 1,2012: http://www.gwec.net/fileadmin/images/Publications/GWEC_annual_market_update_2010_-_2nd_edition_April_2011.pdf.

Goldbrunner, J. (25-29 April 2010). *Austria: Country Update.* Proceedings of the World Geothermal Congress 2010, Bali, Indonesia.

Grace, R., Rickerson, W., Porter, K., DeCesaro, J., Corfee, K., Wingate, M. & Lesser, J. (November 2008). *Exploring Feed-In Tariffs for California: Feed-In Tariff Design and Implementation Issues and Options.* CEC-300-2008-003-F. Sacramento, CA: California Energy Commission. Accessed May 1, 2012: http://www.energy.ca.gov/2008publications/CEC-300-2008-003/CEC-300-2008-003-F.PDF.

Green, B. D. & Nix, R. G. (2006). *Geothermal - The Energy Under our Feet: Geothermal Resource Estimates for the United States.* NREL/TP-840-40665. Golden, CO: National Renewable Energy Laboratory. Accessed May 22, 2012: www1.eere.energy.gov/geothermal/pdfs/40665.pdf.

Guey-Lee, L. (1999). *Renewable Electricity Purchases: History and Recent Developments in Renewable Energy 1998: Issues and Trends.* DOE/EIA-0628(98). Washington, DC: U.S. Energy Information Administration. Accessed May 1, 2012: ftp://ftp.eia.doe.gov/features/renewele.pdf.

Hance, C. N. (August 2005). *Factors Affecting Costs of Geothermal Power Development.* Washington, DC: Geothermal Energy Association. Prepared for the U.S. Department of Energy. Accessed May 1, 2012: http://www.geo-energy.org/reports/Factors%20Affecting%20Cost%20of%20Geothermal%20Power%20Develop ment%20-%20August%202005.pdf.

Hempling, S., Elefant, C., Cory, K. & Porter, K. (January 2010). *Renewable Energy Prices in State-Level Feed-In Tariffs: Federal Law Constraints and Possible Solutions.* NREL/TP-6A2-47408. Golden, CO: National Renewable Energy Laboratory. Accessed May 1, 2012: http://www.nrel.gov/docs/fy10osti/47408.pdf.

Hidalgo, R. (10 November 2011). *Geothermal Potential Studies in Spain.* Proceedings of the GEOELEC Second Regional Data Compilation Workshop. Accessed May 1, 2012: http://www.geoelec.eu/wp-content/uploads/2011/11/R_Hidalgo-Petratherm-Espana.pdf.

Hirsch, R. F. (1999). *Power Loss: The Origins of Deregulation and Restructuring in the American Electric Utility System.* Boston, MA: MIT Press.

Holm, A., Blodgett, L., Jennejohn, D. & Gawell, K. (May 2010). *Geothermal Energy: International Market Update.* Washington, DC: Geothermal Energy Association. Accessed May 1, 2012: http://www.geo- energy.org/pdf/reports/gea_international_market_report_final_may_2010.pdf.

Hunt, J. L. (September 1981). *Geopressured-Geothermal Energy Resource Evaluation for the Northern Gulf of Mexico Basin: An Overview.* BLM

Open File Report 81-01. Washington, DC: U.S. Department of the Interior, Bureau of Land Management. Accessed May 1, 2012: http://www.gomr.boemre.gov/PI/PDFImages/ESPIS/3/3991.pdf.

Hurlbut, D. (2012). *Geothermal Power and Interconnection: The Economics of Getting to Market.* NREL/TP-6A20-54192. Golden, CO: National Renewable Energy Laboratory.

IGA. (2011). "Interactive Map." Bochum, Germany: International Geothermal Association. Accessed November 10, 2011: http://www.geothermal-energy.org/105,interactive_map.html.

Jacobs, D. (2012). *Policy Convergence in the European Union: The Case of Feed-In Tariffs in Germany, Spain and France*. Farnham, UK: Ashgate Publishing.

Jennejohn, D. (April 2011). *Annual Geothermal Power Production and Development Report: April 2011.* Washington, DC: Geothermal Energy Association. Accessed May 1, 2012: http://geo-energy.org/pdf/reports/April2011AnnualUSGeothermalPowerProductionandDevelopmentReport.pdf.

Klein, A., Pfluger, B., Held, A., Ragwitz, M. & Resch, G. (2008). *Evaluation for Different Feed-In Tariff Design Options: Best Practice Paper of the International Feed-In Tariff Cooperation* (2nd Ed.). Karlsruhe, Germany and Laxenburg, Austria: Fraunhofer Institut für Systemtechnik und Innovationsforschung and Vienna University of Technology Energy Economics Group.

Klein, J. (2009). *Comparative Costs of California Central Station Electricity Generation Technologies.* CEC-200-2009-017-SD. Sacramento, CA: California Energy Commission. Accessed June 14, 2012: http://www.energy.ca.gov/2009publications/CEC-200-2009-017/CEC-200-2009-017-SD.PDF.

Kolarevic, V. (30 November 2009). *Geothermal Opportunities and Policy in Serbia.* Proceedings of the Geothermal Energy in South-East Europe Workshop. Accessed May 1, 2012: http://www.ahkungarn.hu/fileadmin/ahk_ungarn/Dokumente/Bereich_HF/Dienstleistungen/Kooperationsboersen/Infoveranstaltung_Kolarevic1.pdf.

Kreycik, C., Couture, T. D. & Cory, K. S. (2011). *Innovative Feed-In Tariff Designs That Limit Policy Costs.* Golden, CO: National Renewable Energy Laboratory. Accessed May 1, 2012: http://www.nrel.gov/docs/fy11osti/50225.pdf.

Lipp, J. (May 2007). "Lessons for Effective Renewable Electricity Policy from Denmark, Germany and the United Kingdom." *Energy Policy* (*35*, 11),

5481-5495. Accessed May 1, 2012: http://www.sciencedirect.com/science/article/pii/S0301421507002091.

Lovekin, J. & Pletka, R. (2009). "Geothermal Assessment as Part of California's Renewable Energy Transmission Initiative." *Geothermal Research Council Transactions*, (*33*), 1013-1018. Accessed May 1, 2012: http://repository.usgin.org/sites/default/files/dlio/files/2011/u19/geothermal_assessment_as_part_of_californias_renewable_energy_transmission_initiative-_james_lovekin_and_ryan_pletka_grc_2009.pdf.

Massachusetts Institute of Technology (MIT). (2006). "The future of geothermal energy: Impact of Enhanced Geothermal Systems (EGS) on the United States in the 21st century." INL/EXT-06-11746. Cambridge, MA: Prepared for the Idaho National Laboratory and the U.S. Department of Energy, accessed August, 20, 2012 at http://www1.eere.energy.gov/geothermal/pdfs/future_geo_energy.pdf

McCrone, A., Usher, E., Sonntag-O'Brien, V., Moslener, U., Andreas, J. G. & Gruening, C. (2011). *Global Trends in Renewable Energy Investment 2011: Analysis of Trends and Issues in the Financing of Renewable Energy*. Frankfurt, Germany: Frankfurt School-UNEP Collaborating Centre for Climate & Sustainable Energy. Accessed May 1, 2012: http://www.unep.org/publications/contents/pub_details_search.asp?ID=6226.

McKenna, J. R. & Blackwell, D. D. (2005). "Geothermal Electric Power from Hydrocarbon Fields." *Geothermal Research Council Transactions*, (*29*), 283-288.

Miethling, B. (2011). "Different but Similar: Geothermal Energy and the Role of Politics in Germany, Iceland and the United States." *Zeitschrift für Energiewirtschaft*, *35*(4), 287-296.

Miethling, B. (2012). *Politische Triebkräfte der Innovation: Eine Analyse der Rolle von Politik in Innovationssystemen der Geothermie*. Frankfurt, Germany: Peter Lang.

Mendonça, M., Jacobs, D. & Sovacool, B. (2009). *Powering the Green Economy: The Feed-In Tariff Handbook*. London: Earthscan.

Mines, G. L., Bruton, C. J., Nix, R. G., Butcher, T. A., Kukacka, L. E. & Premuzic, E. T. (2010). *A History of Geothermal Energy Research and Development in the United States: Energy Conversion 1976-2006*. Washington, DC: U.S. Department of Energy, Energy Efficiency & Renewable Energy, Geothermal Technologies Program. Accessed May 22, 2012: http://www1.eere.energy.gov/geothermal/pdfs/geothermal_history_4_conversion.pdf.

Ministry of Energy. (January 2010). *Feed-In-Tariffs Policy on Wind, Biomass, Small-Hydro, Geothermal, Biogas and Solar Resource Generated Electricity* (1st Revision). Accessed May 1, 2012: http://www.energy.go.ke/wp-content/uploads/2010/08/feed%20in%20tarriffs.pdf.

MEMR. (2011). *The Regulation of the Minister of Energy and Mineral Resources Number 02 of 2011 Regarding Assignment to PT. Perusahaan Listrik Negara (PERSERO) as the State-Owned Electricity Utility to Purchase Electricity from Geothermal Electrical Power Plant and the Benchmark Price of Electricity Purchase by PT. Perusahaan Listrik Negara (PERSERO) as the State-Owned Electricity Utility from Geothermal Electrical Power Plant.* Jakarta, Indonesia: Ministry of Energy and Mineral Resources of the Republic of Indonesia.

NREL. (2011). *Policymakers' Guidebook for Geothermal Electricity Generation.* NREL/BR-6A20-49476. Golden, CO: National Renewable Energy Laboratory. Accessed May 1, 2012: http://www.nrel.gov/docs/fy11osti/49476.pdf.

NSUARB. (April 2011). *In the Matter of the Electricity Act and in the Matter of a Hearing to Determine Renewable Energy Community Based Feed-In Tariffs.* Halifax, NS: Nova Scotia Utility and Review Board. Accessed May 1, 2012: http://www.nsuarb.ca/images/stories/pdf/notices_of_hearing/10Dec/hearing%20and%20order%20-%20setting%20timetable%20renewable%20energy.pdf.

O'Donnell, C., Baumstark, P., Nibler, V., Corfee, K. & Sullivan, K. (2009). *Renewable Energy Cost of Generation Update.* CEC-500-2009-084. Sacramento, CA: California Energy Commission. Accessed June 14, 2012: http://www.energy.ca.gov/2009publications/CEC-500-2009-084/CEC-500-2009-084.PDF.

Ontario Power Authority. (July 2011). *Feed-In Tariff Program - FIT Rules* (Version 1.5.1). Toronto, ON: Ontario Power Authority. Accessed May 1, 2012: http://www.alternativetechnology.ca/uploads/media/FIT_Rules_Version_1.5.1.pdf.

Pierce, K. G. & Livesay, B. J. (1994). *A Study of Geothermal Drilling and the Production of Electricity from Geothermal Energy.* SAND92-1728 / UC-256. Albuquerque, NM and Livermore, CA: Sandia National Laboratories. Accessed June 14, 2012: http://www.osti.gov/bridge/servlets/purl/10131301/10131301.pdf.

PT Castlerock Consulting. (December 2010). *Phase 1 Report: Review & Analysis of Prevailing Geothermal Policies, Regulations and Costs* (Version 1.0). Jakarta, Indonesia: Prepared for the Ministry of Energy and

Mineral Resources. Accessed May 1, 2012: http://www.castlerockasia.com/download/Phase%201%20Geothermal%20Pricing%20Report.pdf.

REN21. (2011). *Renewables 2011 Global Status Report*. Paris, France: REN21 Secretariat. Accessed May 1, 2012: http://www.ren21.net/Portals/97/documents/GSR/REN21_GSR2011.pdf.

Renewable Energy Finance Tracking Initiative (REFTI). (2011). *Aggregate and Renewable Energy Financing Terms and Trend Data- First Half of 2011*. Golden, CO: National Renewable Energy Laboratory. Accessed March 2012: https://financere.nrel.gov/finance/REFTI.

Richter, A. (October 2009). *United States Geothermal Energy Market Report*. Reykjavík, Iceland: Íslandsbanki. Accessed May 1, 2012: http://skjol.islandsbanki.is/servlet/file/store156/item47220/2009%20US%20Geothermal%20Rep ort.pdf.

Rickerson, W., Bennhold, F. & Bradbury, J. (May 2008). *Feed-In Tariffs and Renewable Energy in the USA: A Policy Update*. Raleigh, NC; Washington, DC; and Hamburg, Germany: North Carolina Solar Center, Heinrich Böll Foundation North America, and the World Future Council. Accessed May 1, 2012: http://www.wind-works.org/FeedLaws/USA/Feed-in_Tariffs_and_Renewable_Energy_in_the_USA_-_a_Policy_Update.pdf.

Rickerson, W., Flynn, H., Hanley, C., Jacobs, D. & Solano-Peralta, M. (2010). *Fiscal and Non- Fiscal Incentives for Adopting Renewable Energy: Feed-In Tariffs in Latin America and the Caribbean*. Washington, DC: Inter-American Development Bank.

Rickerson, W., Hanley, C., Flynn, H. & Karcher, M. (10-11 February 2011). *Feed-In Tariff Design: Implications for Financing*. Proceedings of the PV Rollout 2nd European American Solar Deployment Conference, Boston, MA.

Rogers, J., Fink, S. & Porter, K. (July 2010). *Examples of Wind Energy Curtailment Practices*. NREL/SR-550-48737. Golden, CO: National Renewable Energy Laboratory. Accessed May 1, 2012: http://www.nrel.gov/docs/fy10osti/48737.pdf.

Salmon, J. P., Meurice, J., Wobus, N., Stern, F. & Duaime, M. (March 2011). *Guidebook to Geothermal Power Finance*. NREL/SR-6A20-49391. Golden, CO: National Renewable Energy Laboratory. Accessed May 1, 2012: http://www.nrel.gov/docs/fy11osti/49391.pdf.

Sanchez-Guzman, J. & Garcia-de-la-Noceda, C. (25-29 April 2010). *The Evolution of Geothermal Energy in Spain: Country Update (2005-2009)*. Proceedings of the World Geothermal Congress 2010, Bali, Indonesia.

Sison-Lebrilla, E. & Tiangco, V. (2005). *Geothermal Strategic Value Analysis in Support of the 2005 Integrated Energy Policy Report.* CEC-500-2005-105-SD. Sacramento, CA: California Energy Commission. Accessed June 14, 2012: http://www.energy.ca.gov/2005publications/CEC-500-2005-105/CEC-500-2005-105-SD.PDF.

Tringas, T. (7 April 2011). *How Much More Does Clean Energy Really Cost?* Presentation at the Bloomberg New Energy Finance Summit 2011 Roundtable Day, New York, NY.

Tweedie, A. & Doris, E. (July 2011). *Comparing Germany's and California's Interconnection Processes for PV Systems.* NREL/TP-7A30-51814. Golden, CO: National Renewable Energy Laboratory. Accessed May 1, 2012: http://www.nrel.gov/docs/fy11osti/51814.pdf.

U.S. EIA. (November 2011). *Electric Power Annual 2010.* (Table 2.1.B. Net generation by selected renewables by type of producer, 1999 through 2010). Washington, DC: Energy Information Administration. Accessed May 1, 2012: http://205.254.135.7/electricity/annual/pdf/epa.pdf.

Varadarajan, U., Nelson, D., Pierpoint, B. & Herve-Mignucci, M. (October 2011). "The Impacts of Policy on the Financing of Renewable Projects: A Case Study Analysis." Climate Policy Initiative. http:// climatepolicyinitiative.org/wp-content/uploads/2011/12/Policy-Impacts-on- Financing-of-Renewables.pdf.

Williams, C. F., Reed, M. J., Mariner, R. H., DeAngelo, J. & Galanis Jr., S. P. (2008). *Assessment of Moderate- and High-Temperature Geothermal Resources in the United States.* Fact Sheet 2008–3082. Washington, DC: U.S. Department of the Interior, U.S. Geological Survey. Accessed May 1, 2012: http://pubs.usgs.gov/fs/2008/3082/pdf/fs2008-3082.pdf.

Wiser, R., O'Connell, R., Bolinger, M., Grace, R. & Pletka, R. (January 2006). *Building a "Margin of Safety" into Renewable Energy Procurements: A Review of Experience with Contract Failure.* CEC-300-2006-004. Sacramento, CA: California Energy Commission. Accessed May 1, 2012: http://www.energy.ca.gov/2006publications/CEC-300-2006-004/CEC-300-2006-004.PDF.

End Notes

[1] In 2010, a single 15 MW project had been developed in the United States and was installed in Nevada (Jennejohn 2011).

[2] This study does not include a comprehensive look at policies that address exploration risks – just those that could be addressed in a FIT policy; another NREL study (underway in 2012) is examining specific policies that address exploration risks, specifically.

[3] In order to create the conditions for geothermal deployment, the perspectives of multiple stakeholders will need to be taken into account and appropriately balanced. These include, for example, end users and ratepayers, policymakers, and industry in addition to investors. A full multi-stakeholder analysis is beyond this scope of this report, however.

[4] This report does not consider geothermal heat pumps; it concentrates on geothermal electricity production.

[5] The Cost of Renewable Energy Spreadsheet Tool (CREST) can be found at https://financere.nrel.gov/finance/CREST

[6] A more detailed exploration of geothermal technology trends and market development can be found in other publications (e.g., Mines et al. 2010; GEA 2012; Beerepoot 2011).

[7] A number of studies discuss a broad range of FIT design issues, options, and considerations in detail (Couture et al. 2010; Grace et al. 2008).

[8] This figure excludes corporate and government research and development spending.

[9] In addition to these three resources, there are also geopressured geothermal reservoirs that consist of deposits of pressurized hot fluid that typically contain dissolved methane. Geopressured geothermal resources can potentially be harnessed to produce energy from both the heat and hydraulic pressure of the reservoir. The methane can also be recovered and used for energy. Total geopressured resources are unknown, but early studies have estimated that there may be 10-40 GW in the Gulf of Mexico alone (Hunt, 1981).

[10] Hot water dominated hydrothermal resources are far more common that steam dominated hydrothermal resources.

[11] The German Energy Agency defines EGS as "The use of deep heat reservoirs with few or no water resources...Crystalline and dense sedimentary rock at depths of three to six kilometres with high temperatures (over 150 °C) can serve as reservoirs. These are accessed via two or more boreholes drilled deep into solid rock. Hydraulic and chemical stimulation processes (Enhanced Geothermal Systems, EGS) are used to make cracks and fissures in the rock. Cold water is then pumped at high pressure down an injection well into the rock, where it is heated and returns to the surface via a second borehole. This hot water in turn heats a working fluid with a low boiling point (so-called Kalina Cycle and Organic Rankine Cycle, ORC), producing steam for a turbine. Heat can also be fed into district heating networks via a heat exchanger" (German DENA 2012). The U.S. Department of Energy's definition of EGS is not that different: "Enhanced geothermal systems (EGS), also sometimes called engineered geothermal systems, offer great potential for dramatically expanding the use of geothermal energy. The EGS concept is to extract heat by creating a subsurface fracture system to which water can be added through injection wells. Creating an enhanced, or engineered, geothermal system requires improving the natural permeability of rock. Rocks are permeable due to minute fractures and pore spaces between mineral grains. Injected water is heated by contact with the rock and returns to the surface through production wells, as in naturally occurring hydrothermal systems. EGS are reservoirs created to improve the economics of resources without adequate water and/or permeability" (DOE 2012).

[12] This estimate is limited to 11 states in the western United States and to the geothermal resource located at depths of between 3 and 6 km.

[13] Costs that can diverge from estimates made at the stage of competing for a PPA include capital costs, operating costs, fuel costs, costs of securing permits, engineering, or costs associated with PPA pursuit.

[14] Geothermal projects that drill new wells to expand the utilization at existing, well-quantified sites are the least likely to have dry wells because the resource is well characterized (however there likely will be other issues like more quickly depleting the overall resource that feeds both plants). Geothermal projects that "step-out" and are adjacent to existing sites are more likely to have dry wells than co-located plants since the resource is nearby, but

may not continue outside of the existing plant's footprint. Rapid depletion of the shared resources at both sites may also be an issue. Next, greenfield wells at locations with surface manifestations appear to have a higher probability of dry wells than "step-out" facilities since the resource is not characterized at the required depths. Finally, "blind" geothermal that is not near other facilities and does not have a surface manifestation are the most likely to have dry wells (DOE, Geothermal Technologies Program 2011).

[15] Some of these barriers and risks described in this report, such as the construction, technology risks, and resource risks are difficult to address through policy (Corfee et al. 2010; de Jager and Rathmann 2008). Construction and technology risks can be addressed through contracts and warranties with installers and manufacturers, whereas resource risks can be addressed through in-depth resource assessments.

[16] Recent studies have argued that multiple policy instruments that take into account specific national technology and market developments must be combined in order to successfully support geothermal development (e.g., Miethling 2012).

[17] The production tax credit, for example, has often been extended for one to two years at a time, insufficient to span geothermal project development timelines.

[18] Many projects fail not because they do not secure a PPA, but because a PPA at a price bid is insufficient to get the project financed (i.e., to give investors their expected return).

[19] Policies involving utility long-term contract procurement requirements, such as RPS mandates placed upon regulated utilities in western United States markets, can mitigate electricity market risk, but by failing to ensure market access, they expose developers to most aspects of development risk and contract price risk. FITs offering guaranteed access at known prices under standard contract terms can substantially reduce these aspects of power project risk. In addition, such competitive procurement policies fail to avoid many of the costs avoidable under a FIT, such as proposal preparation, cost estimation, hedging to lock in costs underpinning a bid (when possible), risk premium, bid security, and legal and other costs relating to contract negotiation.

[20] Other approaches that include geothermal specific policies and carve-outs, improvements to existing policies, and agency coordination are outside of the scope of this analysis.

[21] Clean Local Energy Accessible Now, see e.g., Caperton et al. (2011) and Farrell (2011)

[22] Similar to a "must take" contract

[23] Similar to a "take or pay" contract

[24] E.g., Gainesville, Florida; Sacramento, California; and San Antonio, Texas.

[25] It is important to note that some of the FIT design options described in Section 3.1 are not currently feasible in the United States. Priority interconnection and dispatch rules, which are a feature of European FITs, are not currently allowed in the United States because FERC requires that interconnection service and access to transmission be non-discriminatory for projects over which it has jurisdiction (Fink et al. 2010). Groups such as the Regulatory Assistance Project (2010) have, however, suggested that priority interconnection and dispatch be introduced in the United States. And, these design features can be used by international readers of this report.

[26] According to PURPA, a small power production facility "means a facility which is an eligible solar, wind, waste, or geothermal facility...and has a power production capacity which, together with any other facilities located at the same site…is not greater than 80 megawatts." See http://uscode.house.gov/download/pls/16C12.txt.

[27] There has also been controversy regarding the payment levels under some FITs, particularly with regard to FITs for photovoltaic systems in countries such as Spain (Couture 2011).

[28] http://www.res-legal.de/en.html

[29] Several countries have policies that are sometimes referred to as FITs but that set a ceiling price for renewable generation, rather than a floor price. Kenya, for example, sets a ceiling price of 8.5 cents/kWh for geothermal (Ministry of Energy 2010). The contract rate must then be negotiated on a case-by-case basis. In Indonesia, generators must compete for

power purchase agreements and the price of the power purchase agreements cannot exceed $0.097/kWh (MEMR 2011). These are not considered FIT policies for this analysis.

[30] The Blumau and Altheim plants receive the Austrian FIT. The 200 kW Simbach-Braunau plant, which is a joint Bavarian-Austrian project on the border with Germany, receives the German FIT (personal communication, Dr. Johan Goldbrunner November 21, 2011).

[31] Italy's current geothermal capacity is not attributable to the FIT. Instead, Italy's capacity was primarily built by the (previously) state-owned enterprise, Enel, which assumed the risk for the entire development process from exploration to power plant construction. It is unlikely that the FIT will be utilized in the near future because current projects in the pipeline are above 1 MW (GeothermEx 2010; Holm et al. 2010).

[32] National Renewable Energy Action Plans were also consulted for each country. Available from: http://ec.europa.eu/energy/renewables/transparency_platform/action_plan_en.htm.

[33] A 3-MW single-flash unit was installed in Qingshui in 1981 and a 300 kW unit was later installed at the same field. Both systems ceased operations in 1994 (Bertani 2005).

[34] Assuming a 7% discount rate and a 72% net capacity factor.

[35] The net present value of the total payments for one kilowatt-hour generated each year under the Czech FIT is €1.91 whereas the net present value of the total payments for one kilowatt-hour generated each year under the Croatian FIT is €1.43.

[36] All of France's overseas departments, with the exception of French Guiana, which is on the mainland of South America, are islands.

[37] Conversely, none of the countries ranked in the top 10 for total geothermal capacity has a geothermal FIT in place except for Italy, whose FIT targets only generators 1 MW and under. In order, the top 10 countries include United States, Philippines, Indonesia, Mexico, Italy, New Zealand, Iceland, Japan, El Salvador, and Kenya.

[38] Available for download at https://financere.nrel.gov/finance/content/CREST-model

[39] Conservative FIT rates are rates that are set at a level to encourage the lowest cost projects. Aggressive FIT rates are set to enable a broader range of systems to enter into the market (Grace et al. 2008). Policymakers can choose to set their FIT rates along the spectrum from conservative to aggressive rates, depending on their policy objectives (Gifford et al. 2011).

[40] The Erneuerbare-Energien-Gesetz or EEG

[41] A degression schedule reduces the fixed rate available to new generators built at a later date as opposed to reducing the rate available to a particular generator over time.

[42] *Bundesministerium für Umwelt Naturschutz und Reaktorsicherheit*

[43] The 230 kW Neustadt-Glewe plant, which was Germany's first geothermal power installation, utilized an Organic Rankine Cycle to generate electricity from lower-temperature water. The project added a power plant in 2003 to an existing geothermal heat plant. The cost of the power plant installation was €1.6 million (GEOFAR 2010).

[44] In almost all cases, the initial drilling cost assumptions from previous years had been exceeded by 70% (BMU 2011a). In order to further reduce drilling risks, the German government also created a grant program to support drilling parallel with the adjustments to the FIT.

[45] Hydrothermal exploration is ongoing in the Canary Islands (Hidalgo 2011).

[46] In New Mexico, carve-outs exist for solar and wind and geothermal shares a combined target with biomass, certain hydropower facilities, and selected other renewables (Doris et al. 2009).

[47] If carefully structured, the relationship between FITs and RPS targets may help mitigate concerns about whether states have the authority to set FIT rates. According to federal law, states do not have jurisdiction to set premium payment rates for wholesale power. States do, however, have authority under PURPA to set rates based on avoided cost. Recent guidance from FERC seems to indicate that state targets for renewable electricity can allow PURPA rates to be set at the avoided cost of meeting those targets. To the extent that states have set a technology-specific target, avoided cost can be defined as the cost of that specific technology for purposes of meeting the target (Cory 2011; Elefant 2011; Hempling et al. 2010). RPS laws do not set payment rates for specific technologies, but they can set targets

for specific technologies. PURPA allows states to set rates, but not above avoided cost. By combining RPS technology-specific targets (under state law) with PURPA rate setting authority (under federal law), states could achieve results very similar to setting technology-specific FIT rates,but only for projects that register as qualifying facilities (QF) under PURPA. This arrangement may not apply for utilities or regional transmission organizations (RTO) that are exempt from signing new PURPA contracts, i.e., a utility would have to voluntarily agree to sign such a FIT/RPS contract.

[48] When offering higher payments for smaller projects, it is important to design a FIT to avoid gaming such as carving a larger project into smaller phases to access higher payments than would be merited by the combined project scale.

[49] It is also important that the eligible geothermal technologies be explicitly defined if it is the intent of policymakers to support a specific type of plant. For example, policymakers may wish to specify which geothermal electric configurations are eligible (e.g., hydrothermal, EGS, co-produced, and geopressured) and perhaps include or exclude geothermal heat or heat pumps.

[50] In order to support this approach, such supplemental policies would need to be confirmed or developed in tandem with the FIT and their relationship formally linked.

[51] A third option is the "spot market gap" structure in which the generator is guaranteed a fixed revenue and paid the difference between that fixed revenue (akin to the strike price in a contract-for-differences) and the wholesale market price over time (Couture and Gagnon 2010). This approach creates a similar risk profile, from an investor perspective, to a fixed, all-in price structure.

[52] There are also different ways to structure fixed price payments. Offshore wind payments in Germany, for example, are designed with higher payment levels upfront and lower payment levels in later years in order to "front load" project revenues. This enables projects to more quickly pay off their debt (Couture et al. 2010).

[53] I.e., the FIT rate calculation presumes revenue post-contract rather than recovering all up-front investment during the term of the FIT.

[54] Different types of investors may have different perspectives on whether shorter or longer-term durations are preferable. Some investors may wish to realize their expected return in a shorter period of time whereas other investors (e.g., pension funds) may wish to invest in contracts with long-term revenues in order to better match their long-term liabilities.

[55] There is a tradeoff when considering caps, queues, and security deposits. On the one hand, creating a "high bar" for entering the queue by establishing high security deposits, etc., could better ensure that only "serious" projects will enter the queue. On the other hand, high barriers to entry may shut out or deter smaller developers from participating in the market. This barrier to entry may constrain the achievement of specific economic development or industrial objectives (e.g., the encouragement of local developers or the deployment of innovative new technologies).

[56] Different technologies, for example, may have different technological risk profiles. Technological risk, however, should be reflected in the financial assumptions utilized to set the cost based rate and should therefore be reflected in the rate itself.

[57] This analysis and the associated model are intended to be used primarily by policymakers considering FITs to support geothermal electricity development. In addition, this analysis and model may also be used by stakeholders in the policymaking process, including geothermal developers, financiers, and utilities.

[58] A 20-year FIT is assumed in this analysis. Sensitivities are also included that test the change in LCOE for FITs of 10, 15, 25, and 30 years.

[59] The term 'aggressive' in this usage connotes a FIT rate intended to support a range of viable geothermal installations, while a 'conservative' FIT would only support the most cost-effective geothermal installations.

[60] The exploration and confirmation phases are assumed to be 100% equity funded and does not include a provision for debt financing.

[61] Geothermal market participants do not universally agree on the distinction of exploration and confirmation costs, particularly as it relates to deep exploratory drilling.

[62] A recent NREL study provides a detailed discussion of modeling inputs and their context (Gifford et al. 2011).

[63] Non-drilling exploration costs are held constant for the purpose of this analysis.

[64] Assumes illustrative 50 MW project. Per literature review (see Appendix A), exploration assumes 25% success rate, confirmation assumes 60% success rate, and all costs assume economies of scale associated with a 50 MW project.

[65] A federal Investment Tax Credit of up to 30% of eligible expenses is available to geothermal electricity projects coming into service on or before December 31, 2012. Thereafter, the geothermal ITC is expected to revert to its historic level of 10%.

[66] DSCR stands for debt service coverage ratio.

[67] Importantly, investors may not make the same assumption. Many investors —particularly debt lenders—prefer to assume revenues only exist during the length of the PPA and thus require that financing term be shorter than PPA term (to protect themselves so they are fully paid back even if the project underperforms in the early years).

[68] The selected sensitivities included herein are intended to represent those variables most likely to respond to the presence of a FIT. Sensitivities to other variables may yield additional insights about the drivers of geothermal power, as a result of factors other than FITs. For example, geothermal LCOE will be highly sensitive to large variations in capacity factor. Annual production, however, is not expected to vary significantly based on the presence of a FIT.

[69] Generation from many PURPA plants was purchased under "must take" provision, which meant that the utility was obligated to purchase everything that a generator could produce but that generators were exempt from having to provide operating reserves and ancillary capacity services (Hurlbut 2012).

INDEX

E

F

G

R

U

V

W

Y